Y0-CBG-197

RESCUE PILOT

DAN McKINNON

McGraw-Hill

New York   Chicago   San Francisco   Lisbon   London   Madrid
Mexico City   Milan   New Delhi   San Juan   Seoul
Singapore   Sidney   Toronto

Cataloging-in-Publication Data is on file with the Library of Congress

McGraw-Hill

A Division of The McGraw-Hill Companies

Copyright © 2002 by Dan McKinnon. All rights reserved. Printed in the United States of America. Except as permitted under the United States Copyright Act of 1976, no part of this publication may be reproduced or distributed in any form or by any means, or stored in a data base or retrieval system, without the prior written permission of the publisher.

2 3 4 5 6 7 8 9 0 DOC/DOC 0 9 8 7 6 5 4 3 2

ISBN 0-07-139119-3

The sponsoring editor for this book was Shelley Ingram Carr, the editing supervisor was David E. Fogarty, and the production supervisor was Pamela A. Pelton. It was set in Janson text per the AV3(mod.) design by Vicki A. Hunt of McGraw-Hill Professional's Hightstown, N.J., composition unit.

Printed and bound by R. R. Donnelley & Sons Company.

 This book is printed on recycled, acid-free paper containing a minimum of 50% recycled de-inked fiber.

McGraw-Hill books are available at special quantity discounts to use as premiums and sales promotions, or for use in corporate training programs. For more information, please write to the Director of Special Sales, Professional Publishing, McGraw-Hill, Two Penn Plaza, New York, NY 10121-2298. Or contact your local bookstore.

Contents

Helicopter Pilots Are Different

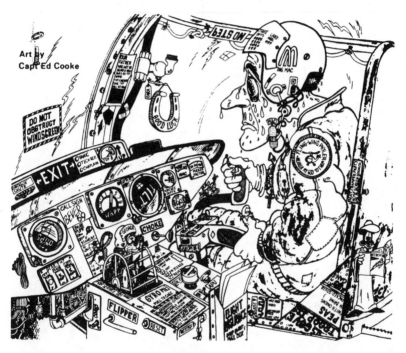

Art by Capt Ed Cooke

"The thing is, helicopters are different from planes. An airplane by its nature wants to fly, and if not interfered with too strongly by unusual events or by a deliberately incompetent pilot, it will fly. A helicopter does not want to fly. It is maintained in the air by a variety of forces and controls working in opposition to each other, and if there is any disturbance in this delicate balance the helicopter stops flying, immediately and disastrously.

"There is no such thing as a gliding helicopter.

"This is why being a helicopter pilot is so different from being an airplane pilot, and why, in general, airplane pilots are open, clear-eyed, bouyant extroverts, and helicopter pilots are brooders, introspective anticipators of trouble. They know if something bad has not happened, it is about to."

—*Harry Reasoner*

Courtesy of CAPT Bill Stuyvesant, USN (Ret.)

Foreword

RESCUE PILOT is a fascinating treatise on helicopter flight, naval shipboard operations, lines of demarcation between the blackshoes and brownshoes of the U.S. Navy, and an incident-by-incident account of the U.S. Navy in the Formosa Strait confrontation preventing the Red Chinese from taking Quemoy.

I enjoyed this book and believe that it has a real place in helicopter history.

Despite their shortcomings, the early visual flight rules (VFRs) for helicopters with piston engines provide guidelines, not all yet recognized, for the needed technical improvements to a naval helicopter.

As the designer and constructor of the first helicopter designed for the U.S. Navy and U.S. Coast Guard rescue operations, the HRP-1, and the HUP, the author notes that these points are well appreciated and serve as a reminder for designs yet to come.

Technically explicit and historically sequential, the story weaves these data into the experience of a young, brand-new, naval aviator with insight as to the need of the trials and efforts toward the serious development of a naval aviator.

Operational risks in shipboard operations are recounted to add to the realism of the story.

The personal confrontations experienced between higher authorities, flying mates, and shipmates, provide background experience for fledgling naval officers.

Adding to these are the seagoing limitations of rough water, winds, and ship motion that describe the real need for vertical-lift aircraft in naval operations, and their immense value to ship operations—not only in rescues, but in ship-to-ship replenishments, underway transfers, pickups, mine detection, and radar calibrations, as well.

The embryonic helicopter shows its potential to physically integrate fleet operations. For the benefit of the nontechnical reader, McKinnon relates the myriad of technical equipment and functions in a story (narrative) manner, not the usual tables of technical data.

Today the grandson of the HUP, the CH-46 *Sea Knight*, designed by Piasecki engineers, continues in these functions trailblazed by the HUP. The HUP life spanned from 1945 to 1970. The HUP and the CH-46 were conceived at the same period in 1945. However, the CH-46 was not built until shaft turbine engines were available in 1958 and is still in service with the Navy and Marines, having served in Vietnam, Panama, the Gulf War, and other missions 36 years later.

The superhuman sacrifice by personnel and their families comes through strong in the story. I salute Dan McKinnon and his fellow shipmates and their families who stand at the forefront of our nation. May the helicopters of the future help them stand firm.

FRANK PIASECKI
President, Piasecki Aircraft Corporation

Acknowledgments

No ONE IS AN ISLAND UNTO ONESELF, and writing a book is a true test of that concept.

Sharing my adventures as a rescue pilot was made possible with help from good friend and aviation expert Dick Palowski with his editing ideas and support, former HU-1 pilot and helo legend Bill Stuyvesant, my Navy flight instructors at NAS Pensacola, and the staff at HU-1, especially chief A. J. Kress.

One person I was especially grateful to was Navy Survival School commander and instructor LCDR (Lieutenant Commander) Bob Chilton, who taught me some of the most valuable lessons in life about myself and others. He taught those same lessons to some 35,000 other naval air crewmen during his career. There's not one of us who doesn't remember him and appreciate his lessons.

The men aboard the USS *Helena* and especially my long-cruise crewmen first class J. D. Dunlap, second class R. A. Karasinski, third-class R. S. Wilson, and second class D. R. Pobst, who all helped make me look good while on cruise with a great maintenance effort.

My genuine appreciation goes to Ken Gazzola, publisher of *Aviation Week Magazine*, for his advice, wisdom, and counsel in making this book possible.

Special thanks to my two secretaries, Patryce Ercolano and Andrea Cooper, for their tireless efforts to whip this book into reading condition with their typing and spelling corrections.

I'd also like to thank Frank Piasecki for designing and building a helicopter I came to love and that taught me the finer points of helicopter operations aboard ship at sea.

My wife Janice was always full of encouragement and constructive criticism as I put the book together.

Most of all, I'd like to thank my great country for the opportunity to serve it in the cause of freedom as a naval aviator.

DAN McKINNON

Introduction

THIS IS NOT JUST A STORY of helo (helicopter) rescue pilots. It also describes how unproven machines, introduced too soon, extracted a toll in blood.

It is also a story of real men and their commitment, character, and sacrifice.

It's a story not told before, one that focuses around the rescue helicopter pilot.

Throughout its history, naval aviation has undergone continuous change. One of the most revolutionary periods of change was the mid to late 1950s when carrier flight operations transitioned from straight-wing jets to the higher-performing swept-wing supersonic fighters. This was a risky evolutionary period with experimental new concepts and aircraft designs that were often plagued with inadequacies and fraught with danger.

It was also a period when the entire carrier force changed to angled decks, steam catapults, and mirror landing systems.

During this evolving period, numerous types of aircraft were introduced, the decks were crowded, at-sea periods long, and the availability of land-based divert fields limited or nonexistent because of unfriendly governments. High-performance jets had little in common with their propeller-driven deckmates. It was an accident-prone environment.

The dangers were illustrated with simple statistics—statistics indicating that the accident rate was the highest in U.S. Navy history. The jets of that era helped develop the concepts that today have provided the United States with the safest fighter, attack, and helicopter aircraft in use anywhere in the world. That progress came at a terrific materiel cost and sacrifice of life.

Statistics show that the present-day carrier-based jets are flying about 60 percent of the hours today compared to those flown in the latter 1950s but with a dramatically reduced accident rates.

Modern helicopters are now flying approximately double the hours with at least 95 percent reduction in the number of accidents and about 20 percent fewer fatalities, despite the fact that today's Navy and Marine helicopters carry more passengers than did the rudimentary contraptions of the 1950s.

It was a unique time to observe this transition in naval aviation as a helicopter rescue pilot.

My primary job and that of my squadronmates was to pluck from danger the unfortunate pilots and individuals in trouble. To accomplish that, many of us faced the same perils. We also had to operate at night in bad weather from storm-blown decks. Many rescue squadronmates throughout the U.S. Navy also paid the ultimate price themselves in the quest to provide a strong defense for freedom.

Unconstrained flying and limited procedural training, coupled with lack of aerial supervision and discipline in those days, spawned many dangers and tragedies. Because there were no guidelines or regulations, naval aviators did a great deal of improvising and, in effect, flat-hatting.

	Carrier jets destroyed	Carrier jet fatalities	Annual hours flown	Fatal Accident Rate per 100,000 hours flown
1956	599	148	790,399	75.78
1957	530	141	854,039	62.06
1958	541	123	868,279	62.31
1959	461	120	822,485	56.05
1997	5	10	513,768	0.97
1998	7	26	494,090	1.42
1999	4	7	484,785	0.83
2000	3	6	422,865	0.71

	Helos destroyed in crashes	Helo fatalities	Annual hours flown	Fatal accident rate per 100,000 hours flown
1956	97	22	165,295	58.68
1957	103	11	199,070	51.74
1958	108	13	236,884	45.59
1959	105	20	264,107	39.76
1997	5	15	385,663	1.30
1998	6	18	408,966	1.47
1999	2	6	408,443	.49
2000	2	11	380,762	.53

Such an appalling mishap rate inspired strict standardization procedures adopted during the early 1960s. NATOPS (Naval Air Training and Operating Procedures Standardization—a program ensuring normal and emergency procedures standardization) established a more professional approach that has consistently led to safer air operations and greater standardization of procedures. The result was dramatically lower accident rates and greater survival rates for Navy pilots. Harsh lessons learned from such delicate handling fighters, such as the F7U *Cutlass* and F3H *Demon*, helped produce safer and better replacement aircraft.

This story is about Navy flight operations that took place before safety became effective and when pilots had the freedom to do and fly pretty much whatever, whenever, and however they pleased while operating their aircraft.

Because of the changing world environments that resulted in the rapid movement of the carrier taskforce, pilots were required to invent their operational procedures as they went along. The one thread that kept the entire thing going was the common bond of their Naval Wings and heritage. But most of all, the pilots had only each other, and the greater risks required of the fighter and attack pilots were accompanied by the greater risks required of the rescue pilots.

It is also the story of one man's experiences and his efforts to save the lives of others and his successes and failures in that endeavor.

RESCUE
PILOT

First Rescue

"ROGER BALL," THE FIGHTER PILOT checked in with the LSO (landing signal officer) on the radio as the F8U *Crusader* turned final approach for a carrier landing aboard the USS *Kearsarge* (CVA-33).

It was a sunny afternoon off the southern California coast with brisk winds in April of 1958.

The jet was about 800 feet above the ocean's surface. The pilot was concentrating on the mirror landing system located on the edge of the port side of his ship near the stern.

This concave mirror with an artificial horizon of green reference lights featured a "meatball" when centered and was adjusted so that the approaching pilot could fly downhill on an imaginary line to the flight deck where the tailhook would snag one of four arresting wires.

But now, the plane was slightly low.

The LSO, standing on a platform near the mirror, was reassuringly coaching the pilot over the radio. He tried to establish a working relationship between himself and the approaching pilot.

"Power, 106," the LSO announced over the radio in a calm voice.

Providing a reverse visual reference of the F8U's position on the invisible 3-degree glideslope, the bobbling red, green, and amber speed lights on the nose gear strut of the F8U, along with lots of experience, told the LSO that this pilot was getting erratic.

If the LSO saw the red lights, this indicated that the plane was flying too fast, green meant too slow, and amber signified that the plane was on speed as it came down the glideslope.

The F8U continued its approach low, below the glideslope. There was no noticeable response to the LSO's call for more power, which would help the jet rise.

"Power," the LSO demanded again with authority and urgency in his voice.

The movable, pivoting wing on the new F8U *Crusader* just entering the fleet was causing lots of consternation for the pilots of VF-154 as they made their qualifying carrier landing approaches.

The fighter pilots were having difficulty adjusting to the fact that they couldn't pull back on the F8U's stick and raise the nose on a landing approach. The idea was to impact the deck on all landing gear at the same time, not flair to catch the wire. High-performance jets were learning to come aboard in a "controlled crash," a concept that is still with us.

Instead of pulling the nose up slightly as the aircraft would settle toward the flight deck, the leading edge of the wing was hydraulically raised up to change the angle of attack while the pilot sat in the same level position regardless of the attitude to horizon. It was thrust, "power," not nose position, that would raise and lower the heavy jets up and down the glideslope. Pilots had to rethink the game, and this was costly.

The approaching F8U pilot belatedly added too much throttle and smartly climbed up through the glideslope and was suddenly high, and prone to a "bolter," or a hook miss.

Obviously his concentration was breaking down. Carrier landings are intensely demanding work.

"Ease off 106," the LSO commanded, but the plane was too close to the carrier and the call came too late.

The F8U slammed down onto the flight deck. The pilot instantly added 100 percent power while hoping to catch a wire. The plane touched down too long on the canted deck. The hook missed the last wire as it banged against the steel deck. At full power with the afterburner torching out the tailpipe, the F8U and its young pilot thundered off the angled deck to circle for another attempt to get aboard.

"106, F8U bolter," the enlisted-man scorekeeper in Pri-fly (control tower) high above the flight deck announced. He used a black grease pencil to mark the unsuccessful F8U "pass" down for all in the glass cage to see on a large plastic wall-mounted tote board.

"Damn, will we ever get this guy on deck?" the air boss announced in a disgusted tone to no one in particular.

It was frustrating to everyone on the giant ship to see a new pilot struggle to qualify as a carrier jet pilot. There was a thin line between rooting support for a young aviator trying to master techniques to get aboard "the boat" tempered by an unspoken disdain for any pilot who just didn't have what it takes.

Our fledgling aviator from VF-154, the *Black Knights* squadron, was going to try again.

He checked in with his fuel state, "106 downwind, 2100 pounds" indicating that he had just enough fuel for a final landing attempt before having to "bingo," or return to Miramar Naval Air Station to refuel. In this case during training workups, the carrier was conveniently located offshore near a major naval air station.

The arresting gear tension of the "crossdeck pendant" (arresting cable) was adjusted for the current weight of the *Crusader*. Too tight an adjustment, and the tailhook could be yanked clear out of the aircraft. Too loose a tension on the arresting cable, and the plane could continue right over the front edge of the canted flight deck before stopping, even after a successful hook engagement.

The *Crusader* lined up astern of the ship again, for his final attempt. The F8U was slightly high this time, but drifting down on the glideslope. Close in to the ship, he cut power and sank ever so low again.

"Slightly low, give me just a little power," urged the LSO.

The F8U banged onto the deck.

The landing gear shock absorbers nearly bottomed out from the force of the impact. The black-and-white striped tailhook snagged the number 2 wire as the arresting cable played out and the jet fighter came to a sudden stop in less than 200 feet.

High above the flight deck in Pri-fly, the air boss looked out his octa-gon-shaped birdcage of thick tinted-glass windows, swiveled his high-back, black padded-leather seat to his right, depressed the lever on the old-fashioned-looking metal squawk box and called:

"Bridge. Pri-fly."

"Go ahead, Pri-fly."

"We've got him trapped, Cap'n."

"I thought we'd never get him aboard. We're about to run out of steaming room. Let's get him taxied out of the landing area and tied down so that we can turn downwind and recover the helo."

"Aye, aye, Sir."

"Who was that kid? Will he be ready tonight?"

"He's a nugget named Peterson, Cap'n, I'll check with the LSO to see if he's ready for night quals."

The man in charge of the carrier control tower turned to his assistant, rotated his chair back to view the flight deck 50 feet below, picked up the flight deck 5MC PA (public address system) mike and with a firm deep voice ordered:

"Taxi 106 forward. On the double. Get some chains on him so that we can turn downwind."

The *Crusader* had been reeled backward by the retraction of a 2-inch-thick arresting cable, and the pilot had raised his tailhook after it released the arresting cable. The plane was now free to taxi forward, and the pilot added power to move ahead. He followed the rapid and precise arm direc-tions from a yellowshirted flight deck crewman. As he signaled, the crew-man leaned back into the 35-knot wind blowing across the flight deck.

"106 was the last qual for this batch, Sir. Ten apiece for all five pilots in the daisy chain" (continuous circling of pilots making landing attempts), the air boss announced to the ship's captain.

"I hope we can get 'em six landings apiece tonight. Some of those guys are a little raggedy," the captain answered back from the squawk box. "I count three bolters for that kid in 106."

"Hold on a sec, Sir, let me check with the LSO and see where we stand with him."

The air boss grabbed a handset phone, "Paddles. Pri-fly."

"Go ahead, Pri-fly."

"Good job on 106."

"He was a challenge!"

"Can he hack it tonight?"

"I don't know, I've got to talk to him. Let me get back to you."

"Roger."

"Well, we've got to be ready for the second launch of the afternoon in about 30 minutes," the air boss told everyone in Pri-fly, knowing that they'd pass the word on to all concerned. Now it was time for a quick break in flight ops (operations) aboard the carrier, affectionately known as the "Bird Farm."

And thus the 28,000th landing aboard the USS *Kearsarge* was safely accomplished. A cake had been baked to honor the occasion, although the pilot couldn't be there to enjoy it. He was busy in the ready room reviewing his poor performance with the LSO. Elsewhere, sailors on the flight and hangar decks would enjoy the sweets and celebrate their contributions to the record.

The giant carrier heeled to starboard as the captain ordered her into a left turn. The *Kearsarge*, an Essex class carrier and veteran of the Korean War, crisply responded in the choppy Pacific off the shores of southern California. A fresh salty sea breeze swept across the flight deck generated by the ship's 25-knot speed. As soon as the ship turned downwind, its speed would neutralize the wind, making it safe to recover her planeguard helicopter.

Several brownshirts and blueshirts led by a yellowshirt leaned forward as they plowed their way through the wind across the flight deck. They were carrying wheel chocks and tiedowns to secure the helo once it landed.

"Stand by to recover the helo," the air boss announced over the flight deck loudspeaker.

"Angel, this is Wildcat, I'm turning downwind. Stand by to recover," the air boss called me, using both the ship and my call signs.

"Roger, Wildcat," I answered matter-of-factly through my lip mike attached to my flight helmet.

"Angel, say your fuel state."

"Forty minutes to red light."

I had 40 minutes before a mandatory requirement to land. When the bright red light on my instrument panel illuminated, the aircraft handbook said that we had 30 minutes' flying time or 100 pounds of fuel left before the tank went dry. But it was a very primitive and unreliable gauge. All through training we'd been taught that you must land soon after the red light came on. Some pilots had engine failures from fuel starvation long before the advertised 30 minutes.

As helicopter rescue pilots, we had consistently tried to impress on the ship's air boss that if he wanted a helicopter available to rescue his jet jockeys, he had to get us back aboard for gas when the red light came on, *no matter what.*

"Pri-fly, this is the captain."

"Go ahead, Cap'n."

"Is the helo pilot part of the training group, too?"

"That's affirmative, Cap'n. He's a "jg" (junior grade) named McKinnon. This is his first flight aboard *Kearsarge.*"

I'd been flying my HUP-2 planeguard helicopter sideways, keeping my relative position along the starboard (right) side of the carrier up by the bow for catapult shots or back near the stern for landings. I was doing this while flying 100 feet above the water for the past hour. Flying sideways at 25 knots required constant attention and control movement using both arms and legs for flight control inputs to adjust to the buffeting effects of the wind.

The planeguard's job was to pluck from the water any unfortunate pilot who had an underpowered "cold cat" shot and dribbled off the front of the carrier at such a low speed to force him into the water instead of flying. I would also be there to rescue any person overboard, or anyone who crashed while learning to land his jet on the carrier. In any kind of accident, my job was to fish the aircrewmen, with the help of my determined expert rescue crewman, out of the unfriendly ocean. My helo was always launched before fixed-wing flight operations got under way and we were invariably the last aircraft recovered.

During daytime, the most common type of carrier accident was a "ramp strike" on landing because of problems maintaining the tricky glideslope. This means hitting the back of the ship.

If the pilot was lucky, a ramp strike would only shear off the jet's landing gear. The jet would then skid along the flight deck, trying to get airborne. Since the landing gear was hydraulically actuated, all the flight control hydraulic fluids would drain out of the aircraft in about 30 seconds. That would cause the aircraft to go out of control.

A pilot had less than 30 seconds to eject, if he was to do so, while still having control of the aircraft. That was assuming that the pilot was fortunate enough to hit the top of the ramp or the flight deck.

If, however, he hit the bottom of the ramp, he'd go into the "spud locker," blow up, and cause a gigantic fire on the fantail of the ship.

It was dangerous work for us all, especially because we were all fresh from flight training and new to shipboard operations. We had to qualify with a lot of preparation ashore before going out to the carrier and have our skills tested at sea in real life.

We were all Navy pilots, which meant that we all had to learn to operate our aircraft off ships. No matter what kind of aircraft we flew, each one had its own set of problems that required close pilot attention.

I sat strapped into the left seat of my helicopter with the trap door and rescue cable and hoist beside me to the right as I watched the jets land. My crewman, who operated the rescue hoist and would pull in the pilot being rescued, sat belted to an aluminum-framed canvas-covered bench behind me.

I soon found that my grip on the collective (control handle that caused the helo to climb) and the cyclic controls worked in empathy with each novice carrier pilot. If he was high on his approach, I found myself nearly flying into the water as I would unconsciously push the collective down, lowering me toward the sea as I mentally tried to help the fixed-wing pilot down so that he could catch an arresting cable.

If he was coming in low, the next thing I'd notice was that I'd be up around 300 to 350 feet trying to help lift him up from hitting the spud locker. One potential problem of getting too high could be a midair collision as a section (two) or division (four) of jets would come smoking up from the stern on the starboard side of the ship in preparation for the carrier break to enter the landing pattern.

From the water surface up to 200 feet was my private airspace. I was never supposed to go higher, and in those days, as today, you never wanted to get into a disagreement with the air boss over space.

This was my second week aboard a carrier. Several weeks previously, I'd spent a week aboard the USS *Bennington* (CVA-20), a sister ship of the *Kearsarge*, which had been in on the kill of the giant Japanese battleship *Yamato* in 1945.

I was beginning to feel pretty comfortable aboard one of these iron monsters handling my tandem rotor helicopter and flying sideways about 200 yards from the edge of the flight deck.

This gave me a ringside seat to watch all the carrier action. Whatever might happen, I could maneuver my helo just as if I wore it on my back to watch any incident. I was even getting good at judging which were the best carrier jet pilots.

Once the last jet had been recovered, I left the planeguard position and started flying the chopper in large circles instead of sideways to get the feeling of the controls' response in normal flying conditions forward to prepare for the landing.

It had been an uneventful flight. No problem with the jets. I was at the age where the squadron could give me some bread and water, an old mattress to sleep on, plus my helo, and I was the happiest guy in the world.

So far I had made no rescues. But I was constantly gaining in shipboard experience. Actually, for all the hazards involved in carrier aviation, it was relatively safe when things worked smoothly. The training officer back in my squadron, HU-1, had said, on the average, that only about 25 percent of the guys in our rescue squadron ever got a single rescue during their career. Most of the time there would be a lot of challenging flying, but rescues were rare. It turns out that rescues are like MiG (Russian fighter aircraft named after Mikhail Gurevich) kills for aces; for some reason only a small percentage get most of them.

After all the aircraft had been recovered aboard the *Kearsarge*, the ship turned downwind to bring me aboard. Intensive flight operations had been going on all day, although this was my first flight. This wasn't training only for me. Three other pilots from the squadron were there to learn how to operate off the ship. We had to learn to find our way around this steel city maze called an *aircraft carrier.*

We were led by one experienced pilot who was charged with the responsibility of making sure that we all survived and didn't interfere with the carrier's flight operations or schedule. The helo detachment was expected to be seen, but not heard—ready in case it was needed, but not in the way. Those jets sucked up fuel fast. They didn't want some stupid helo pilot screwing up their operations. But again, the carrier strived to become one well-oiled machine. Everything and everyone had a place if it was to work efficiently.

This was the first day of this one-week at-sea period. As soon as I landed aboard the ship the plan was to shut down the helo and refuel, and that I'd stay with it while waiting for the launch of the second batch of qualifying jet pilots later in the afternoon.

Our HUP-2 was so primitive that we had no instrumentation to keep our balance safely in flight during night ops over the water. So we stored it with blades folded in a corner of the hangar deck during night flight operations. Any rescues would have to be by lifeboat after dark.

At that time, any pilot in the water at night would depend on a plane-guard destroyer following in the carrier's wake. Powerful shipboard searchlights were used to help spot the pilot, hopefully, floating in the water. Then the destroyer would launch a small whaleboat to pull him to safety. This technique had a poor record of success, but it was the only method available until a better helo was produced.

The ship was slowly turning downwind as the huge wake frothed and churned the choppy Pacific.

Once steaming downwind, the relative gust across the deck would drop from a howling 35 knots to nearly calm. This would make it easy for me to come alongside the ship and establish the helo in a hover in a relative position to the ship. Then I would gradually slide the helo from over the water, across the flight deck, at about 10 feet of altitude. I would then land, following the directions of my crew chief. After the helo had been secured to the deck using the tiedowns, the delicate hollow rotor blades could be stopped without gusts of wind slamming them into the side of the helicopter. This would cause them to break off and fly into anything nearby. If my landing was at the end of flight operations, the blades would be folded so that the helo would take up less valuable room. During ops, a canvas boot would be slipped over the end of each blade and tied to the helo to prevent the blade tips from being caught by a wind gust and snapping the blade in two.

While flying I had two radios tuned in my helicopter. One was for the ship's control tower frequency and the other, the universal emergency "guard channel" frequency, which enabled me to hear anyone broadcasting a distress call.

As the *Kearsarge* continued to turn, my thoughts about landing were interrupted as the guard channel in my helmet suddenly crackled.

"Mayday, Mayday! This is (garbled) ejecting southwest of (garbled)."

I alerted like a hunting dog.

Who is he, and where is he, I thought as my adrenaline started to flow. Naturally I wanted to make a rescue. Maybe this was my chance. This was my job. I depressed the mike button on my cyclic control.

"Wildcat, this is Angel. If he's near here, I'll go after him," I volunteered.

"Stand by, Angel."

"Pri-fly, bridge," the speaker barked.

"Roger, we heard it bridge. We'll try to get him on guard."

"Bridge, CIC [Combat Information Center]. We got the IFF [identification, friend or foe] emergency transponder squawking. Bearing at 345 True, 15 miles. We're tracking it."

"Roger."

Then the radio transmission echoed through the radio speaker on the bridge.

"Aircraft in distress. Aircraft in distress. This is the USS *Kearsarge*," the air boss radioed on 243.0 UHF, the emergency frequency in an effort to get the troubled pilot to answer.

"This is North American test flight 7. I'm at 8000 feet with a flameout and I'm ejecting when I reach 5000 feet. I'm east of Catalina Island near Avalon Harbor."

"This is Long Beach Coast Guard. Can we be of assistance?"

The radio waves became garbled with static as everyone within radio distance tried to offer assistance.

Finally the *Kearsarge* commanded the airwaves. "This is the USS *Kearsarge*, we're taking charge of the rescue. Everyone on guard frequency remain silent. Long Beach Coast Guard, we're sending our helicopter."

"Angel, this is Wildcat; go get him. Your magnetic heading is north."

"Roger. I'm on my way."

All my training back at Ream Field flashed through my mind.

"Hang on, Suarez," I told my crewman. "Let's see if we can find the guy who ejected. Be sure the horse collar is ready to go down the hatch. Help me look out the canopy to see if you can see a raft or anything."

Suarez had heard all the radio calls, too.

"The hoist is ready, and I'll keep looking, Sir." He unfastened his seatbelt, crawled forward, kneeling over the closed rescue hatch, and started looking through the canopy.

I turned right 90 degrees, headed north, and twisted the motorcycle-type grip at the end of the collective with my left hand to keep the blade rpm [speed (revolutions per minute)] from slowing down as I lifted the collective to add power and gain altitude. Simultaneously I pushed forward on the cyclic. The 16-foot-long rotor blades dug into the air as we climbed from 200 feet on up to 500 feet to get a better view in the clean sunny weather to look for the downed pilot. The HUP cruises best at 65 knots of airspeed.

But I was going for my first rescue. I thought to myself "this is no time for normal cruise."

I kept the collective higher than normal and pushed the cyclic forward. The airspeed moved up to 80 knots. I wanted to get there as fast as possible. Today's rescue helo can reach over 200 knots, but for the HUP, 80 knots was tricky.

Vibration throughout the entire airframe and controls increased. The vibration in my seat spread throughout my body trunk, arms, and legs. It scared me a little when the nose wanted to pitch up. To keep from tearing the machine apart, I had to back off on the power and slow down to 75 knots.

The magnetic compass wobbled around in the water bowl instrument housing as I pushed in the right rudder slightly and moved the cyclic toward the right. The water compass swiveled and tilted from the helicopter vibrations, but the lubber's line generally indicated north.

This was my only navigation device in this primitive second-generation contraption with whirling rotor blades on the top. No fancy gyroscopic devices in the cockpit. Nothing more sophisticated to navigate with than a compass similar to that used by a hiker in the woods. At least I was on course and the ship could track me on radar. They provided occasional vectors to the position last indicated on their radar.

The ship lost contact with the troubled aircraft when it was 15 miles away. It was heading west and merged with the south end of Catalina.

As I continued my northward journey, time seemed to stand still. I tried to mentally review all my training, but was distracted by the need to watch airspeed and keep a constant lookout for a raft, flare, parachute, dye marker, or someone swimming in the water. I didn't know what to expect to see. My actions seemed to be automatic, reflecting just what I had learned in training.

Suarez continued to quietly peer out the front right canopy, intently scanning the water. We just kept vibrating as we flew. We didn't talk over the intercom. We knew our jobs. Then off to the left I saw a column of black smoke rising from the top of the southernmost mountain ridge of Catalina Island.

"Was that fire a trash dump or a plane crash?" I wondered.

As a youngster, I'd sailed with my dad to Avalon Harbor on the southeast side of Catalina more than a dozen times and never remembered any dump burning on top of the rugged mountain. It must be the plane.

"Did the pilot get out?" I wondered to myself. "He radioed that he was going to eject, but did he?"

"Angel, we show that you have 5 more miles to go."

"That's about 4 more minutes," I thought.

"I see black smoke on top of the island, but nothing in the water," I radioed, straining to see anything.

The carrier was now steaming toward me to close the gap between me and their last position, which would make for a quicker return flight, assuming that I made a rescue and the victim needed immediate medical attention.

"Suarez, test the hoist and be sure that the sling is attached properly."

I wanted to be sure that the thing worked right when we got there. "Have you got the comealong on?" I inquired. A "comealong" is a vice grip device that attached to the hoist cable. It was worn on a strap that went around the crewman's shoulders and under his armpits to enable him to attach himself to the cable. Once he clamped onto the cable, I would then use my thumb to operate the hoist with a switch on the cyclic stick grip. This let me lower my dangling crewman into the water to help an injured pilot get in the rescue sling. Then I would hoist the crewman back into the helo after he attached the downed pilot into the rescue sling. Once in the helo, the crewman would then hoist the rescuee into the helo. We often practiced this awkward maneuver, but it was extremely complicated and dangerous and to be used only as a last-resort technique.

The pilot really couldn't see what was happening in the hatch below him because there was no stabilizing device to hold the helo steady. I had to keep my eyes on the horizon and just steal glimpses of the action below me as I hovered, using both arms and feet.

"I've already tested it, Sir, and I'm ready to go down."

About 3 miles south of the island I noticed a few dots in the distance—pleasure craft and yachts.

Then I spotted it: green dye marker in the water ahead and to my right about a half mile. "That's gotta be him," I thought.

I made an adjustment in my course. Then I started my descent from 500 feet down toward the water's surface.

The HUP could drop like a rock. I eased my collective down, backed off on the power by rotating the motorcycle grip handle inward, and pulled back on the cyclic, slowing the airspeed to 45 knots. My pulse rate accelerated. I started my descent at 1000 feet a minute, down to 200 feet, then down to 50 feet. I disciplined my flying to remember all my training to be sure that there were no screwups. I wanted this rescue. Who knew if I'd ever get another chance. I saw a small one-man yellow life raft on the edge of the dye marker with a white chute deployed nearby under the water.

"Wildcat, there's a green dye marker in the water with a raft nearby. I'm going down."

"Roger."

We were flying cross-wind. Since I sat on the left side of the helo, I moved to the right to keep the downed pilot in sight.

Finally I was abeam the dye marker.

I flew down past the raft and then started a U-turn to the left to come up into the wind just above the water. Heading into the wind would make hovering easier for the pickup.

I continued to slow and started to transition to the hover. The rotor blades slapped the air.

The helo shuddered.

We started to sink toward the water.

I pulled higher on the collective, simultaneously twisting the motorcycle grip to add power. I was now at 37 inches of manifold pressure. The red-line maximum was 42.5 inches. Normal cruise was 30 inches.

The helo settled on the ground-effect cushion—the air compressed by the downward force of the rotor blades against the water.

Salt spray kicked up—but it took less power to hover in the ground effect about 10 feet above the water.

I eased forward on the cyclic, and the helo gradually moved forward—but I was keeping the downed pilot in sight. It took right rudder to keep the helo steady.

"Open the hatch, Suarez," I said over the intercom, "and be sure you drag that horse collar along in the water before we get to him."

I didn't want any sparks or static electricity to affect us or the guy we were about to rescue. We needed to ground the helo to the ocean before the pilot touched the sling.

As I started to slow down and begin the hover, I saw our victim for the first time.

"Who is he, and where did he come from?" I thought.

His chute had red-and-white panels. Navy chutes were all-white. He was wearing a bright orange flight suit. Navy flight suits were a light brown canvas. He had a strange-shaped flotation device, and the biggest pair of black boots I had ever seen on a pilot. Our standard U.S. Navy issue was a gold helmet, a yellow flotation device with "BUAER" stenciled on it, which stood for Bureau of Aeronautics, and brown flight deck boots.

"We'll figure out who he is later," I thought. "Let's just get him fished out now."

I came to a hover about 10 feet above the water and 25 feet from the downed pilot. The downwash of the rotor blades was a powerful force. The pilot shielded his face from the blast to protect it from the force of the salt spray.

Suarez had the trapdoor on the bottom of the helo open and was peering down to find the pilot.

He was dragging the sling in the water.

The horse collar sling is a tricky device. If the rescuee gets in right, it's almost impossible to fall out as he is hoisted into the helo. If he climbs into it backward, he's sure to slip out and fall back into the water when he's about half way up into the helicopter. But most important of all, he needs to be free of his parachute.

There were instances of helicopters picking up a pilot in the water still strapped to his chute. The rotor downdraft filled the attached chute with air, and as it filled, combined with the natural wind, it produced a tug-of-war that the helo could not win. The chute always won the contest by pulling the helo into the water.

If the rescued pilot was quick, he could cut the parachute shroud lines and save everyone. If he failed to do so, the helo crewman could take a pair of bolt cutters located near his seat and clip the rescue cable. It would save the helo if accomplished in time, but the rescuee would be in big trouble. The wind would fill his chute and pull him in the water, separating him from the raft and the helo. Then the helo would no longer have a way to hoist him aboard.

To help, a huge "ABANDON CHUTE" warning was painted on the bottom of the helo near the rescue hatch.

Fortunately, this man had his chute removed. I moved in over him.

This was the tricky part: hovering without moving or being able to look down and see the man. You just hovered right on top of him and picked up a point on the horizon for reference. Your crewman directed you by talking to you over the intercom.

Adrenaline had me pumped up for my first rescue. And if the law of averages was right, it would be my only rescue.

Suarez called on the ICS (the intercom system), "Forward about 3 feet, go right one foot." Salt spray kicked up, and was covering the canopy. I moved the cyclic forward and to the right.

"Steady."

I stole a quick glance down to my right through the open hatch.

In the fresh sea breeze, the raft was dancing around in the water because of all the rotor downwash.

Looking down, I lost reference on the horizon. I almost slipped too far to the right.

"Just concentrate on the hover and let Suarez be your eyes," I thought.

"Come left a little, Mr. McKinnon," Suarez directed.

"Angel, this is Wildcat."

I was too busy to answer the rapid call from the ship.

I slid to the left.

I looked down again.

"He won't leave his raft, so we'll drop the sling in his lap," I thought.

He grabbed for the horse collar. But the raft slid around on the water. He got out of the raft into the water. Now he entered the sling.

"He's getting into the sling," Suarez called over the intercom.

"Steady, Mr. McKinnon."

With total concentration, working the collective with my left hand, cyclic with my right, and with each foot pushing the rudder pedals back and forth to compensate for the wind, I held the helo steady.

"He's in the sling," Suarez said.

"Wow. The way he got into the sling amazed me. Either he's had some great training or he's done this before," I thought.

"Weight coming on the hoist."

"He's coming up, Sir."

As his head came through the hatch with his gloved hands above his head holding onto the hoist cable, Suarez stopped the hoist for a second, took our rescuee's hands, and turned his body so he faced toward the rear of the helo. Then with the pickle switch in one hand and the steadying of the rescued pilot with the other hand, he started the hoist up again. I could only watch out of the corner of my eye.

The pilot in the orange flight suit lifted his legs up on the flooring aft of the rectangular hatch in the floor that he had entered.

This was important. Our squadron lost a helo during a rescue when a downed pilot came facing forward through the hatch and reached for the first handle he could find to help hold himself steady. It just happened to be an octagonal red handle on the control pedestal that was the idle cut-off mixture control! He simply shut off the gas to the engine, and the pilot did a 2-second autorotation into the water. It was impossible for the pilot to get the gas back on before the crash. The helo sank. The pilot and his crewman almost drowned getting the trapped rescued pilot out of the rescue sling before the sinking helicopter pulled them all to a watery grave.

Suarez thought it was easier to close the hatch while our dripping rescuee hung in the sling attached to the hoist overhead. His heels were on the solid floor of the helo toward the rear, and his own rear end was raised above the open hatch that he had just come through. It took about 10 long seconds to secure the hatch so that he had something solid to rest on.

I sure hated to see the chute and raft left behind, but after losing a multi-million-dollar airplane, taking a chance on saving a couple of thousand dollars of rescue equipment that had done its job seemed too high a risk. Besides, a pleasure craft in the area would probably pick them up and return them to someone.

Once the hatch was closed, Suarez reversed the hoist motor downward and lowered the sling to let our rescuee lie on the floor of the helicopter. He then helped our orange-suited visitor squirm out of the rescue sling. Soaking wet and shivering from the cold water, the pilot just sat on the floor.

"Be sure to leave his 'Mae West' on, Suarez."

I pulled up the collective and added power by twisting the throttle grip and then pushed forward on the cyclic. I adjusted my rudder pedals, left the hover, gained transitional lift forward, picked up airspeed, and at 45 knots climbed up to 200 feet—the bottom of the deadman's curve.

Suarez couldn't talk to our rescuee above the engine and rotor noise, so I handed him my kneeboard and told him over the intercom to have the pilot write his name and where he was from.

I adjusted the course to return to the carrier. Suarez handed me back my kneeboard with shaky pencil scrawl.

"No wonder he looked so different," I thought. "Wonder what kind of aircraft he was flying?"

"Wildcat, Angel. We've completed the pickup. We're on our way back to the ship. Our rescuee appears to be in good shape. His name is William R. Yoakley. He's a test pilot for North American Aviation."

"Roger, Angel."

"Get him off the floor and strapped into the seat beside you," I ordered Suarez.

I didn't want anything to happen to him. If we had a problem with the helo and had to ditch ourselves, I wanted him strapped in.

I knew back on the carrier that the skipper would be saying something like, "Get all the aircraft respotted and manned for the next launch. The helo should be here in 6 minutes. We'll recover him amidships. Medical personnel stand by. We're behind schedule for our next launch. We're going to be pressed to get all the traps in before night ops commence."

Navigation on the return trip was easier. I climbed up to 500 feet and spotted the carrier in the distance and I just headed for it. The ship had closed the distance. Right after my radio call I saw them start turning downwind so that the flight deck winds would be light for my recovery.

There was no need to rush now. I had my quarry, and he appeared uninjured. I just returned at normal HUP cruise of 65 knots.

Approaching the ship, I radioed, "Wildcat, Angel, ready to come aboard." This is one time I had the full attention of the entire carrier and knew that there would be no trouble getting landing clearance.

"Angel, land in the clear area by the island." One reason to land here was that it made it easier for medical personnel to get the rescued pilot to sick bay if he needed attention.

"Roger."

The jets were lined up around the flight deck, obviously getting ready for a launch. They were parked by the forward cats and stacked wingtip to wingtip on the fantail.

"Angel, we'll refuel you on the run and launch you as soon as we can."

I thought they would allow for a pilot change, but I guess I'd just have to wait until later to learn about the guy I rescued.

My yellowshirt crew leader had his hands raised over his head like a referee indicating a touchdown as I pulled alongside the ship at about 20 feet above the flight deck. I hovered there momentarily as I changed my technique from flying over water to pick up a relative position alongside the port side of the ship.

I checked the signal flags high atop the ship's island to verify that the relative wind was satisfactory. I also checked that the ship exhaust stack gasses would not cause the helo added turbulence.

Then I gradually eased my helo sideways to my right over the flight deck as the yellowshirt, with his back to the bow of the ship, held his left arm at right angles to his body, and used his right arm to wave me over to the center of the flight deck. When he had me at the spot he wanted, he crossed both arms overhead in a giant X formation and pushed his arms down to indicate for me to land.

I eased the helo down to a landing. The two side-mounted wheels and tailwheel simultaneously touched down. Greenshirt crewmen ran in from each side to put large yellow wooden U-shaped chocks on each tire so that the helo wouldn't roll on the moving flight deck. I had the brakes set, but aboard ship you don't take chances. Then the crewmen attached a tiedown to a circle ring on each side of the helo, and hooked it tight to a flight deck tiedown spot.

I bottomed the collective, but kept the rpm and centrifugal force high on the hollow rotor blades to keep them from coning and snapping off. It would be a real hazard if there were any gusts of wind.

Bending down and hunched over because of the low overhead inside the helicopter, Yoakley came up, shook my hand with a wide grin, and then turned and exited the helo via the aft left door. He was met by medical personnel wearing white shirts with big red crosses and carrying a wire stretcher.

He didn't need any help. The stretcher carrier followed the doctor, who escorted Yoakley past the plexiglass nose of the helo toward a door in the island on his way to sick bay for a physical.

An additional adventure was awaiting Bill Yoakley, too.

Meanwhile, the redshirts were refueling me with the rotors and engine turning. The dangerous procedure known as "hot refueling" was prohibited ashore, but fairly routine aboard ship when time was critical.

I took on 300 pounds of high-octane aviation 115/145 gas, enough for an hour and a half of flying. I saw the redshirt pull the hose back and retreat from the aircraft to the deck edge. It would be years before helos would have turbine engines.

Once they were clear, I nodded my head toward the aircraft director that I was ready for launch and waved my fist with thumb pointing sideways between my legs. He put his fists down with thumbs extended by his knees and pushed them both outward, indicating to each man stationed at the tiedown rings to take the straps off. They did, and then ran forward to him so that I could see each tiedown in their hands and know that I could lift off without the danger of one side of the helo still being attached to the deck.

Another squadron pilot had proved that it was disastrous in a devastating crash with rotor blade shrapnel flying everywhere. At this stage in the history of naval aviation, helicopters had a very high accident rate, although a higher percentage of jet accidents were fatal. Nonetheless, the launch was a risky time, as I was to find out later.

With the tiedowns clear of the helo, I lifted up on the collective and the helo rose straight up 10 feet, out of the chocks, and I maneuvered sideways over the edge of the ship. I circled around as the ship turned into the wind to commence flight operations again. I really felt good now, with the rescue being a real confidence builder.

The jets had started their engines. I could tell by the heat waves coming out of the rear of the aircraft. All I could hear in my cockpit was the constant thrashing of air by my rotor system, the loud engine exhaust, and radio crosstalk in my helmet.

I took my place flying sideways into the wind slightly above the starboard flight deck edge. The "FOX" flag was flying from the yardarm, indicating that we were on course to resume flight ops. Then the first fighter was flung off the catapult, made a slight dip below the flight deck, and safely accelerated forward and upward.

The next hour and a half of planeguard flying was routine. I found myself wondering about my recent rescue of Yoakley.

It hadn't taken too long for me to become a member of the "Pelican Club"—a special recognition for a helo crewman who had saved someone from the water. I was no longer a virgin. And much to my surprise, this was to be just the start of my rescue flying career.

CHAPTER TWO

War Stories

JUST PRIOR TO A SUNSET, which glowed like a fireball as it melted through fading cumulus clouds, the last jet slammed onto the flight deck. Its tailhook snagged an "OK3" arresting wire.

The *Kearsarge* turned downwind, straightened up on its heading, and brought me aboard. Control of the helo was difficult as I flew through the turbulent air caused by stack gas. The hot unstable air caused the chopper to bounce around. It took a lot of effort to keep it under control on landing and while being secured to the flight deck.

I disengaged the rotor blades from the engine and flicked the friction switch on the instrument console to quickly slow the rotor blades before any wind gusts could snap them. My crew swarmed over the helo. They climbed to each rotor head and pulled the T-handles connecting the blades to the articulating dampeners on the rotor head.

Then they put in place long poles with a cradle on top to rest the end of each rotor blade. This was the first step in folding up the rotor system to protect the tips of all the blades. The blades would be held in a fixed position over the top of the helo fuselage before it was lowered on the flight deck elevator to the hangar deck and stored out of the way during night ops.

The crew chief wanted to know what mechanical gripes I had against the aircraft. They would be repaired before anyone got to sleep that night. The helo would also receive a freshwater washdown inside and out to flush all the corrosive salt spray acquired during the rescue hover.

The biggest enemy to a helo based at sea was saltwater corrosion.

I completed the "yellowsheet" discrepancies, listing the mechanical problems for the maintenance guys, as crewmen started to tow my helo backward with a tail wheel tow bar to the deck edge elevator. My flying day was over, but I still had an hour of paperwork to complete, including the rescue report.

I headed below deck toward my cramped stateroom located farthest aft in officers' country. This was an ideal room location and allowed me the quickest access to the helo in case of an emergency launch for "man overboard" or any other crisis. An added bonus was that this area was the smoothest-riding part of the ship and farthest away from the banging of the noisy catapults.

I hung my sweaty flight suit in the closet, took a quick shower in the community head (bathroom) down the passageway, completed my paperwork, and was off to the wardroom. Chow was served continuously in the wardroom while flight ops were taking place.

I arrived at the wardroom during a lull. Those pilots flying the first launch of night ops had already left to suit up and brief. The pilots I had planeguarded for, plus the flight deck and operations officers, were now starting to drift in for the evening meal.

As members of the helo detachment aboard the *Kearsarge*, we really didn't belong to anyone—either the ship's company or the air group. Our squadron had units deployed aboard carriers from Naval Auxiliary Air Station Ream Field near San Diego, California, led by an officer in charge. He was usually a lieutenant, with three junior officers in tow for training purposes. During these short 1- or 2-week training cruises, the same air group with all the squadron planes and maintenance men "worked up" with the same ship over a few months in preparation for a long Far East cruise. It would last 6 or 7 months.

HU-1 used these "workup" cruises to assign different planeguard helo detachments in order to rotate young pilots through the short cruise. They would eventually be assigned to carriers, cruisers, or amphibious ships for long-cruise deployments. The result was that the helo pilot never really got to know anybody on the ship for more than 5 to 10 days. Looking around the wardroom to try to find William R. Yoakley was a fruitless exercise.

Everyone was a stranger to me and vice versa. I didn't see him and finally found a steward who said Yoakley had eaten with the ship's operations officer about half an hour before I arrived. Yoakley was overheard saying that he was stuck spending the night aboard the ship. I had fully expected him to be flown ashore in the COD (carrier on-board delivery) aircraft or the ship's airline to shore).

After dinner I climbed the five stories to the bridge and asked the OOD (officer of the deck) where Yoakley was. They had assigned him a 2-man stateroom forward near the number two catapult. When I reached his assigned stateroom and knocked on the door, it was opened by a man with rugged features, square jaw, curly black hair, and eyes that radiated friendship. He was dressed in a borrowed khaki uniform with no tie and no insignias on the collar. His broad smile showed even with a giant cigar clenched between his teeth.

"Hi, I'm Dan McKinnon. I picked you up this afternoon, I said, thrusting my hand forward.

"Well, I'm glad to meet you. Thanks for the ride," he said with a firm handshake. "Looks like the captain is going to keep me a prisoner here."

"What do you mean?"

"I'm spending the night. They're not going to fly me ashore until tomorrow. I think the Navy wants to make some kind of production out of your rescue."

"What were you flying, anyway?

"An F86D Sabre jet. It had just been overhauled. This was its first flight. I was doing a routine production test flight, and the engine flamed out. I couldn't get it to relight. Guess I'm lucky you happened to be in the area, although it might have been more fun to have been rescued by one of those yachts over at Catalina with all those young, good-looking girls in their bikinis."

"When we got you in the helo I couldn't figure out where you came from with all that strange-looking gear."

"We've got some special equipment that North American Aviation issues to their test pilots. See this life vest we use," he said, as he reached for a yellow rubber jacket from the back of a chair.

"You guys in the Navy have the toggle that punctures a CO_2 cartridge to inflate your Mae West after you enter the water. We've got an extra gadget that clamps onto the CO_2 cartridge holder. It has a plunger. Once water seeps into the extra chamber, it wets a seltzer tablet, which expands, pushing the plunger down to activate the CO_2 cartridge. This inflates the life vest automatically."

"Huh!" I commented. "I wonder why the Navy doesn't have something like that?"

"Don't know, but a little over a year ago it saved one of our test pilot's lives."

Yoakley told me that George Smith was the first pilot to ever eject at a low altitude while going supersonic and lived to tell about it. He was flying an F100A back in February of 1955 when the controls jammed at 35,000 feet.

"The plane nosed over and began a steep dive. He wanted to save the plane. He frantically yanked and pulled at the controls. They were frozen. He continued at a near-vertical dive for 28,500 feet, nearly 5 miles straight down. At 9000 feet he decided to abandon the flawed jet. It had already broken the sound barrier. At 7000 feet, just before entering a cloud bank, he blew the canopy off and a half a second later at 6500 feet, he and the ejection seat left the plane. They were in the clouds now.

"As he left the cockpit, the airblast he entered knocked him unconscious. He decelerated with the force of 23 Gs [23 × gravity constant]. His eyeballs were almost yanked out of their sockets. They turned red from the ruptured blood vessels. His intestines and internal organs were bruised and damaged from the hydraulic forces of the sudden stop when he hit the air as he left the plane.

"The last thing he remembered was his airspeed indicator reading Mach 1.05—1.05 times the speed of sound or 777 miles an hour at that altitude. Two and half seconds after the ejection, he was at 4800 feet and automatically separated from the seat.

"His arms and legs flailed loosely and he was tossed around as though a kid had thrown a Raggedy Anne doll from a rooftop. We have a special device that automatically opens the chute below 10,000 feet. His opened at 4100 feet.

"The panels in the parachute ripped out as it slowed the fall of his body. It was nearly shredded. At 2000 feet he came out of the clouds. The plane had already smashed into the water and was pulverized into pieces the size of cornflakes. He remained unconscious when he slammed into the water off San Pedro. This little device worked as advertised and inflated automatically when it got wet. It saved him from drowning.

"A fishing boat nearby saw the whole thing happen. A former Navy rescue specialist aboard the boat pulled him from the water within 50 seconds, just before he was about to drown in his tangled parachute.

"He spent 6 months in a hospital and a year later was testing airplanes again."

I got to thinking about how easily Yoakley had entered the horse collar that afternoon.

"You sure got into the rescue sling easy this afternoon. You ever done that before?"

"Yeah, sort of, once."

"Were you rescued before?" I asked.

"Yeah."

"How did it happen?"

"It was wild. I was flying out of Taegu [in South Korea] in F-80s. Remember the *Shooting Star?*"

"That was the first operational jet fighter, wasn't it? Made by Lockheed?"

"Right, but besides 50-caliber machine guns, we had strapped on some rockets. Our mission that day was to head north of Wonson and pick targets of opportunity. There were four of us. The weather really deteriorated the farther north we went. Cumulus buildups were everywhere. We decided to try and stay low under the weather and weaved in and out between the mountaintops.

"Then down in this bowl-shaped valley I saw some enemy troops and vehicles. I called to my wingman, Willy Wall, to orbit overhead while I went down to strafe and fire off a few rockets. After I made my pass, I'd called him to follow me again. I rolled over and 'split-S'd' down toward the center of the valley.

"On the way down, a road carved on the near-vertical side of the mountain about 1000 feet from the valley floor caught my eye. Then I saw some trucks, real targets for a change. They were about 2000 feet below the ridge line.

"I maneuvered to put the row of trucks in front of me like a shooting gallery.

"As I leveled out from the dive and started to fire my 50-caliber machine guns, I had really gained speed—more than 450 knots. It made it impossible for the troops with machine guns on the valley floor to hit me, but it was really fast for this bowl-shaped valley with near-vertical walls.

"By the time I got lined up and aimed for the trucks, it was too late. I wasn't going to be able to pull up and keep from plowing into the mountain dead ahead. I pulled back on the stick. The plane didn't respond. It just seemed to continue straight like a speeding bullet. Finally the movements of the controls took effect and the aircraft started pointing upward.

"I could see the mountaintop just above my nose and wondered how long it would take to pancake into the mountainside with the belly of the aircraft and end up in a ball of fire. What a stupid way to die.

"I yanked back harder—all the way on the stick. It was in my lap.

"I thought I was going to bend the aluminum stick from the pressure of my hand—I was pulling so hard. I knew I was going to stall if I missed the mountain. I was out of the ejection seat envelope and I didn't have time to eject, anyway.

"So all I was buying was about 15 seconds more of life if I cleared the mountaintop and stalled out over the top of the mountains. But I was going to die trying to keep from dying."

I asked Yoakley if he was scared.

"There was no time to be scared. For a second I thought I was going to clear the mountain. Then right in front of me was this crag in the mountainside that jutted out. I ran out of room and struck the edge of the lip.

"The plane ricocheted upward like a projectile out of an exploding gun. My head felt like it had been torn loose from my shoulders. My helmet split against the edge of the cockpit.

"Acid gray clouds of smoke poured into the cockpit. My throat burned from the toxic fumes. I grabbed for my oxygen mask dangling from one side of my helmet. Got it strapped over my face so I could breathe and switched it to 100 percent. That kept the smoke out of my lungs.

"Then I tried to wave my hand to get the smoke away from the instruments so I could see what was happening. At least the gloves would keep my hands from burning even though I thought I was about to die.

"I blinked till my eyes watered. They burned from the smoke. I pulled my goggles down to protect them.

"Fanning the smoke was a waste of time. It was seeping in through every crack and crevice and duct in the cockpit.

"The engine and the accessory compartment right behind me were ablaze. I pushed the stick forward. I wanted to level off. Next I pushed the throttle full forward. I couldn't see anything. The fire was behind me. I knew I was out of the ejection seat envelope. The only thing I could do was to pull the yellow-and-black striped handle to blow off the canopy and clear the smoke so I could see. I felt around for the handle and pulled it. The canopy blew off. The smoke and fumes were sucked out of the cockpit. The perspiration that had soaked my flight suit was now turning to ice. Was I going to freeze to death?

"The wind swirled around the cockpit. It made it harder to keep oriented on what the plane was doing.

"Billows of black smoke trailing the aircraft mushroomed to plumes of ugly black smoke. My wingman, Willy Wall, was shouting over his radio to me. I was too busy to respond.

"He waited tensely for the fireball explosion that would destroy my aircraft and me along with it. The heat on my shoulder blades was getting hotter, but somehow I was flying. I wondered what I was flying. I squeezed the transmitter button on the stick and radioed to Willy.

" 'Gas Mask Baker Leader. Can you hear me, Willy?'

" 'Yeah, I can hear you, Bill. For crying out loud, get out of that wreck.

" 'Come in and check me.'

"Willy was counting the seconds until the whole thing disintegrated and radioed to me.

" 'Get the hell out of it fast.'

"Even though the plane was about to fall apart, I ignored Willy's advice.

" 'Let's head south to Wonsan," ' I radioed.

" 'South?" ' he questioned back. 'Are you nuts? It's at least 100 plus miles from here.'

" 'Call the Navy,' I responded.

"He did, and lined up a rescue 'copter near Reito Island—friendly territory. I thought if I could hang together long enough to make it out to sea before the F-80 fell apart or blew up, at least I'd have a chance to avoid becoming a POW [prisoner of war].

"The rushing wind was now going through the gaping holes in the fuselage, tearing the plane further apart.

"Then I got this wild idea. If I climbed above the turbulence, the wrenching pressures on the aircraft might disappear and allow the plane to stay together. I slowly climbed to 12,000 feet. Then to 15,000 feet. The air was smooth, but colder. My body started to numb and my feet began to freeze.

"I had a choice. Stay up there and freeze to death or go back down to a lower altitude and let the plane tear apart. I went back down.

"The throttle linkage was gone. It didn't matter what I did with the throttle lever. As I moved it fore and aft, the engine kept working. Fortunately at a constant rpm.

"So I got ready to eject. As long as that F-80 held together, I was going to stay with it.

"Finally I crossed over the shoreline and out to sea. From one enemy to another. I feared a parachute landing in the water. I'd heard many stories about pilots drowning in their chutes. I tightened my seat harness. There were lots of stories about back injuries from ejections.

"Off to my right down below, out of the corner of my eye, I saw a long, lonely bug-looking machine skimming over the water with the sun glistening off the windmill rotor blades.

"There was the helicopter. My helicopter!

"Then there were no more choices. The right wing was shivering, and then it folded up like a broken arm, snapped, and drifted away.

"I went into a steep dive.

"At this instant the six machine guns shorted [short-circuited] out and began firing. I was on my back, then upright as what was left of the plane started to spin violently. The jet engine was still burning with cruise thrust. I was doing at least 500 knots.

"I clutched the ejection handle, pulled it, and catapulted clear of the plane with what felt like a mule kicking me in my seat.

"As I was tumbling in the ejection seat, the plane exploded with a deafening roar, but I was clear of it.

"My hands fumbled around and then I found my lapbelt. I managed to unbuckle it and drifted away from the seat and pulled the ripcord D ring. The chute blossomed open with a sound that sounded like a slap. The risers jerked on me. What a great feeling!

"Then I noticed that some of the shroud lines were tangled over the top of the chute. This caused me to fall faster than normal and gave me no control or ability to steer the parachute.

"I was being carried farther out to sea.

"As I passed through 3000 feet, the helicopter circled around me. I grinned and the pilot waved back.

"For a moment I relaxed from the tension I'd been under and was thankful to be alive. I even thought about home, but I was still cold and realized the freezing water below would allow me only 10 minutes or so before going unconscious. I still had a life-or-death experience ahead of me.

"My fingers found the parachute harness snaps. There were two near my groin and one across my chest. I kicked my feet as I got near the water so I wouldn't release them too early. Others had misjudged their height above the water and had done high-altitude belly flops.

"As I hit the water I got the two fasteners unhooked. I started to sink, then kept going deeper with the weight of all my survival gear and the 200 rounds of ammunition for my pistol. The water pressure in my ears was enormous. All that stuff designed to save me was going to be worthless, I thought, because I was going to drown before I got to use it. My fingers were already starting to get numb from the cold water. Finally, I got the chest buckle located and undone and reached for the two toggles to inflate my Mae West.

"Slowly I came to the surface, but I wasn't free from my parachute. The wind was dragging me along. I started to gag and choke on the salty water.

"I'd done everything right, yet I couldn't get free of the harness. My left leg started hurting. Then I realized the chute harness was fouled around my leg.

"About that time the horse collar was lowered on a cable to me from the helicopter. I couldn't get into it. I just grabbed the sling and held on for dear life.

"The crewman started the hoist, and the tug-of-war was on. I was holding on, but my arms were getting numb from the cold water. The tangled parachute harness on my leg was pulling me back into the sea. It felt like a tourniquet on my leg. The pain in my arms was excruciating. Then the leg strap fell free and the parachute blew away.

"I was going to live if I could just hold on a while longer. I was swinging over the waves as the hoist got me near the helicopter. Then the crewman reached out for me and pulled me into the small cabin. I was freezing, cold, hurt all over, exhausted, but finally safe. All I could say to two of your old squadronmates was 'Thanks, fellows.'

"Aw, don't mention it," said the crewman, a fellow named William Moore.

"No dramatic words, but for me a dramatic life saving event.

"The pilot, Lieutenant Don Whittaker, flew us back to the LST [landing ship, tank] they were operating from. He was stationed offshore just for such an event.

"Then I was transferred to the cruiser USS *Manchester* (CL-83). I was held hostage there for 6 days, much like tonight. The Navy made the Air Force send out a case of Johnny Walker for ransom, before they'd release me.

"I wonder if North American Aviation will have to do the same for your rescue?"

I asked Yoakley, "Wonder what caused your plane to flame out today?"

He replied, "Dunno. But the investigation will probably figure it out.

"From that experience, I realized how important it was to learn the proper technique of entering the rescue sling, and that's why I did it right today."

"Today was a piece of cake compared to your Korean experience."

"It's never easy when you leave your aircraft and depend on your parachute then work to survive in the sea. The water is never forgiving."

I was to find that to be true many times in the next 2 years.

I called it a night, saying, "Well, let's hit the sack. I've got to get up early for the first launch."

I walked back to my stateroom through the maze of hatches and passageways.

I was greeted by the slamming and banging of the catapult working the night launch and bounces. A noise I'd listen to until I fell asleep.

How I got here this particular night all started when I got inspired reading a book while attending the University of Missouri. It was the dramatic and heart-rending naval aviation story by James A. Michener, titled *The Bridges at Toko-Ri.* Somehow I related to all the characters of the book, F9F-2 *Panther* jet fighter pilot Harry Brubacker, HO3S helicopter pilot Mike Forney, and helo crewman Nestor Gamidge.

It was a simple, yet dramatic story about bravery, skill, and courage. What a sad ending, but it motivated me even more to join the Wings of Gold Fraternity.

Then came the movie with William Holden and Mickey Rooney. I relived that fateful story over and over again and again. Now I was positive I was going to fly in the Navy—but fly what? Jets or helicopters? One thing I knew, I was going to be the only pilot on whatever I flew. It had to be a single-engine.

I joined the NROTC (National Reserve Officer Training Corps) program while attending the University of Missouri as my ticket to becoming a naval aviator.

As an NROTC midshipman I had to go on a cruise between my junior and senior years. The U.S. Navy collected hopefuls from all over America and sent us to Norfolk, Virginia, where all midshipmen boarded a fleet of ships for a 6-week cruise. We were issued sailor outfits and sailor hats with a blue stripe on top that branded us as midshipmen—the lowest form of Navy life.

I was assigned to the heavy cruiser USS *Roanoke* (CL-145). We sailed around the Caribbean, and visited Guantánamo Bay and Havana, Cuba and the Panama Canal.

We learned how to navigate by taking star fixes; holystone the decks in unison; clean hot sweaty engine rooms; eat food most people would throw away; and take orders from lowly seamen on our way to becoming officers.

"If you can't take orders, you can't give them," went the slogan.

The *Roanoke* was the flagship of our armada of 14 or so ships with NROTC students, plus Naval Academy midshipmen who thought they were one degree less important than God.

Our ship had the admiral on board, and he was entitled to a helicopter. We had a new HUP-2. Two pilots flew this helo off the fantail nearly every day. I watched in fascination. That poor pilot in charge—I bugged him daily, every free moment, about his helo, about how it worked, about flying, and especially to fly in it with him.

Just to keep me quiet, I'm sure, he finally gave me a flight toward the end of the cruise. What a thrill to take off from that rolling warship and the challenge the pilot faced to come back in that vibrating machine for the tricky landing on a pitching deck with gusty winds.

For the 6 weeks of the cruise, that poor helicopter pilot couldn't get rid of me. I helped his crew polish and clean saltwater spray from the helo and followed him around like a puppy dog. I don't remember his name, but I'd sure like to thank him for his patience.

The blackshoe or seagoing shipboard Navy with its engine rooms and officer of the deck watches wasn't for me. I really wanted to become someone different—the "brownshoe" Navy and aviation was my goal. I was convinced.

My senior year of college was different from the relaxed and fun-filled early years. I was really motivated to fly in the Navy. I had a goal and knew it. My grades greatly improved.

On graduation in June of 1956 as I wore my dress white uniform, it didn't take me long to find an enlisted man to give me my first salute as an officer—and then for me to fork over to him the traditional dollar bill.

No one was happier. I had my orders to Pensacola, and navy flight training starting in September.

But first, since the training pipeline was full, there was to be a 3-month delay at Miramar Naval Air Station—the Navy's master jet airbase in San Diego—waiting for my training slot at Pensacola.

Life is what you make it, and to me orders to Pensacola was the end of one dream and the beginning of another.

But first Miramar was to give me a few scars and sobering incidents to be sure that I wanted to try my turn as a navy pilot.

CHAPTER THREE

Miramar

W HEN I CHECKED IN at Miramar Naval Air Station—the Navy's largest jet fighter base—the base administration officer pushed his khaki hat back and scratched his head, trying to figure out what to do with the lone ensign assigned to him to kill time for 3 months. He had more important things to do than babysit this single would-be aviator who ended up on his doorstep.

His idea was to put me in an obscure office of special services in a corner of the base far away from flying! My job would be to worry about theater tickets, bowling, and ballgames.

After all that effort to get through college, it wasn't my goal to end up as part of some athletic operation in Navy Special Services. I wanted to be a part of flying.

So I shared that with the administration officer. Thankfully he relented and agreed to assign me to the Operations Department.

For a young military enthusiast who dreamed of wearing the Navy Wings of Gold, I felt I had now arrived. The squadron pilots would come over to our building usually with their flight gear on, sometimes including G-suits, helmets, and oxygen masks. All the tools of the trade I longed to possess. I tried to act cool, but enjoyed the cocky way they swaggered around the base. I worked diligently to learn their language and acronyms for the trade of navy flying.

At Base Ops, they would file their cross-country flight plans and check the weather. The FAA (Federal Aviation Administration) traffic control radar was on the second floor and the control tower was perched on top of the building. Visiting fighters and dignitaries parked their planes on "our" ramp.

The officers and chiefs were patient with my enthusiasm and constant questions. They gave me assignments to learn more about how naval aviation worked—how to file flight plans, read detailed weather reports, and stand watches in the control tower. They provided plenty of written rules and regs (regulations) to keep me busy reading.

I also became acquainted with the crash crew and their emergency vehicles parked right next to our tower.

At least two or three times a day, they'd don their silver asbestos fire-fighting suits and roll their red firetrucks and ambulances to line the runway ready to assist some fighter jock who had to make an emergency landing. They were on the cutting edge of excitement and danger.

It happened in the middle of my second week. The reality of flying military jets was introduced to me. I arrived at the Operations building to a quiet, somber mood. It didn't take long to figure out that something was wrong.

Just an hour before there had been a launch of two F7U *Cutlass* fighters. They took off in staggered formation with afterburners torching out the rear of the twin-engine jets. Just as the accelerating planes lifted off the ground, one of the engines on the lead aircraft caught fire and streamed flames.

A red fire-warning light flashed in the cockpit.

The pilot immediately radioed his problem to the tower and tried to climb out in a right-hand pattern to avoid flying over nearby homes.

He was intent on saving his aircraft while not putting anyone on the ground in danger.

The flames spewing out from his tail exhaust expanded and grew brighter.

Climbing and turning downwind, he was intent on circling back for a landing.

It was strange to be in a right-hand traffic pattern when most navy aircraft make all turns to the left.

He left his landing gear and flaps down so that he could land immediately. The F7U resembled a flying wing that flew in a cocked-up position while in slow flight. In that position, it had what appeared to be a 45-degree nose-up angle with a nose gear oleo that hung down and looked as long as a telephone pole. It was a weird-looking flying machine.

The wingman of the pilot in trouble radioed to his leader about the increase of the fire streaming from the rear of the aircraft. He urged the young *Cutlass* pilot to climb higher so that he could safely eject.

In this dangerous jet, you needed speed and altitude to be sure of ejecting in one piece.

It had a poor ejection seat.

Zero-zero ejection seats for a low-level escape hadn't been developed yet.

The pilot continued, intent on salvaging his aircraft. These were terrifying moments before death—too low to eject, yet doomed to hit the ground.

Normal turns into the final approach to the end of the runway were always dangerous in the underpowered *Cutlass*.

But now parts started falling off the plane. The crippled craft disintegrated into flames.

It fell short of the runway, crashing into the officer's club with a full load of fuel, and exploded on impact in a ball of fire. The club was destroyed as giant flames licked the morning sky.

All that was left of the plane was a charred metal skeleton of scattered pieces. The "O [Officers] Club" was a burned-out shell of a building. Fortunately no one was in the club at that early hour.

When told of the crash, I drove my car down to see what had happened. The fires were all out. Investigators were standing around and the smell of burned flesh and chemical firefighting foam, which smelled of pulverized horse dust mixed with water, gave off a stench that burned in my nostrils and which I can still remember. It was a bitter odor you learn to associate with death and tragedy.

I learned firsthand why the *Cutlass* was called the "Ensign Killer."

A couple of years before, U.S. Navy test pilots had tried to convince Pentagon brass not to buy the plane, saying that their test flights convinced them it had very bad handling characteristics in the landing configuration. It was relatively powerful and a true supersonic fighter, but for takeoff and landings it required an above-average pilot. Because politics entered the picture, a lot of good men needlessly lost their lives.

For a day or so the crash dampened my enthusiasm for flying. But flight operations continued. Virtually all the jets were single-seat fighters. A couple of TV-2 (T-33s) dual-seat trainers were committed to training, so it was impossible to bum a ride in them.

Besides, I hadn't been checked out in the pressure chamber and ejection seat trainer, yet.

The pilots in operations had three types of planes assigned for use: (1) an old DC-3, known in the U.S. Navy as an R4D; (2) an F2H Banshee twin-jet straight-wing, single-seat fighter; and (3) a HUP-2 rescue helicopter, similar to the one I had finagled a ride in during my midshipman cruise.

At the time, the U.S. Navy still had a few flying chief petty officers, enlisted pilots [or APs (aviation pilots)] left over from World War II who hadn't retired. They were not in squadrons, but assigned to various air stations. The officers got stuck with most of the paperwork, and it was the chiefs who got to do most of the flying.

We had two of those APs stationed at Miramar and one, Curly Simmons, took a liking to me. He tolerated answering all my machine-gunned questions about navy flying. He knew I had a private pilot license.

One day, Chief Simmons offered me a ride in the HUP. "Here, climb up in the cockpit with me and sit in the right seat. You'll get a better view up here than in the back."

I struggled to climb between the two seats in the cockpit. Boy, was I excited as I tried to outwardly act casual. Awkwardly, I lifted my legs over the control console, then over the collective, and plunked down in the seat. Moving around, I finally looked out the front canopy. "Wow, what a view." That panorama in front beat sitting in the back.

I sure wanted to look as though I knew what was going on to the ground crewman helping to get us under way. Matter of fact, I wanted to look like I knew what was going on to anyone watching. If they thought I was a navy pilot sitting in the cockpit, so much the better.

The chief got the engine cranked up, engaged the rotors, and took off from Miramar and cruised over some of the flatlands and rolling hills north of the field.

While we were scooting along about 100 feet above the terrain, the chief said the magic words, "Why don't you see if you can handle the controls. Put your feet on the rudder pedals and hold the stick with your right hand. I'll still hold onto the collective. The controls are sensitive, so just make small corrections."

"Okay, I've got it."

It wasn't long before we bounced around like a yoyo.

"Take it easy. You don't have to move the stick all around the cockpit—just little movements," Chief Simmons said firmly.

His voice didn't have the alarm in it I thought it should have as he responded to the kind of flying I was doing.

We bounced around all over the sky for a while until he finally got tired of the rough ride and took back the controls.

I was excited to have had the opportunity to try my hand at piloting one of these machines. What a challenge!

Simmons flew us north to Hour Glass Field, an old glider airport and abandoned auto racetrack north of Miramar. The chief demonstrated precise control of the helicopter as he flew the helo sideways with the nose following a straight line along the perimeter of the field. Adroitly, he made turns at the corner of the field with the back of the helo swinging around, so we maintained our nose position perfectly on the edge of the old runway.

Curley was a real master at controlling this contraption.

Then we left the field and started a sight-seeing tour buzzing through the canyons, valleys, and hillsides in the area. We scooted down a remote dirt road about 50 feet off the ground for a couple of miles, chasing jackrabbits and looking at the sagebrush.

The chief spotted a car in the distance parked off to the side of the isolated dirt road. He grinned and abruptly turned the helo to the right and headed for a hilly area out of sight to the people sitting in the car. It took me a moment to catch on. His idea was to come around undetected and approach the car head on.

"Bet we find a little action going on down there. Let's surprise them," he told me over the intercom.

Swiftly he circled around the hill out of view of the occupants of the car. When we approached the car, he pulled back on the cyclic stick, slowing the helo, and brought it to a hover just 20 feet from the hood of the car.

We peered in as the fan effect of the rotors swirled a cloud of dust all over the car and helo. The startled guy in the car peered out the windshield at this ugly intruding helicopter hovering in midair, making a terrible racket just a couple feet above the ground.

He saw two pilots with helmets and dark eyeshields covering their faces peering in on him.

He and his girlfriend didn't know what to do first—roll up the windows to keep the dirt out or cover up. The clothes won.

"Bet that guy is cussing us out a blue streak," the chief laughed. "Normally, they hear you coming with the thump-thump of the rotor blades, but by coming around that hill, we really surprised them."

"We sure did!" That wasn't the last time I pulled this newly learned stunt.

Enough was enough, and the chief finally lifted up on the collective and pushed the cyclic stick forward. We climbed on out, leaving a whirlwind of dust engulfing the lovers' car.

"We'd better get out of here before that guy gets out of the car and starts throwing rocks at us," the chief smiled.

The chief was a great storyteller. When we landed back at Miramar, he had all the guys at Operations laughing so hard their sides hurt the way he told the story.

Stories of the Blue Angels are legendary.

One day I noticed on the arrival board that two of the Blue Angels were inbound to Miramar from their base in Pensacola, Florida. They were going to refuel here on the way to an air show in northern California.

The two blue-and-gold F9F-8 Cougar jets arrived overhead in their tight echelon formation and peeled off at the break with perfect abrupt 180-degree 5-G turns to circle back for a landing. They spaced out their aircraft so that they could land on each side of the runway one just behind and alongside the other.

Then a chill hit Operations.

We monitored one of the pilot's radio calls to the control tower.

"Miramar, this is Blue Angel Four, I have an unsafe nose gear indication. I'm going around for a check."

"Blue Angel Four, Blue Angel One, I'll pull up and check, Nello."

With that, Commander Zeke Cormier, the leader of the Blue Angels and World War II double ace, pulled his F9F-8 Cougar up out of the flight pattern and joined wing on Lieutenant Nello Pierozzi.

"Nello, your main gear is okay, but the gear doors on your nose gear aren't even open. Recycle your gear."

Pierozzi lifted the gear handle, and the main gear was sucked up into the wheel wells.

The two planes climbed to 3000 feet over the airfield. Pierozzi tried twice more to lower all three landing gears—with the same result—the nose gear wouldn't extend.

"Skipper, I'm getting low on fuel. I better go on down. I have enough fuel to make a bounce to see if I can jolt the nose gear loose, then I'll have to make a final."

"Roger, I'll go down first because they'll have to close the field after you touch down."

"Okay, boss."

"Blue Angel Four, would you like us to foam the runway for you?"

"Negative, I don't have time. I'll come on in for a touch and go."

"Roger."

Commander Cormier touched down and taxied off the main runway. The firetrucks, ambulances, cranes, and emergency vehicles raced with red lights flashing and sirens blaring to take their assigned positions along the runway.

The tower closed the field and directed all incoming traffic to North Island Naval Air Station.

The ramp area in front of Operations began to fill up as spectators drifted in. Somehow the word of danger spreads quickly.

Lieutenant Pierozzi came around slowly, made a carrier-style bounce landing, added power, and roared into the air—but the nose gear still didn't extend.

"Tower, it didn't work, I'm coming around for a final. I'll hold the nose off as long as I can."

He circled around the landing pattern again to get into a position for landing, gradually coming down the glideslope.

Now the veteran stunt pilot, who flew the dangerous number 4 slot position in the Blue Angels' acrobatic formations, eased his main gear down ever so gently on the runway, holding his nose high in the air to let the wings act as aerodynamic speed brakes.

Then he lowered the nose of his fighter to a foot or so off the concrete runway. He held it there until the stick came back in his lap. The swept-wing jet ran out of airspeed and lift. The nose gently fell though, scraping along the runway with a shower of red-and-white sparks spewing back toward the underside of the fuselage.

Pierozzi immediately applied his brakes, and the plane stopped quickly. The emergency vehicles raced toward the crippled plane.

Within seconds, the aircraft was surrounded by foam-laden fire trucks and silver-suited men ready to spray the plane and pilot in case flames burst out.

Pierozzi opened his canopy, unstrapped his harness and chute, and quickly climbed out of the cockpit. Rapidly he walked away from the crippled fighter. Nothing happened. No smoke. No fire.

The plane's nose, with its skinned-up landing gear door, was lifted onto a wheeled flatbed and towed to the Operations Building.

Now the real excitement started.

As leader of the Blues, the short-tempered Cormier had a schedule to meet, and he told the top man in Operations that he wanted the damaged jet fixed and fixed fast. The lieutenant commander in charge remarked that he thought they "could have it fixed by sometime tomorrow afternoon."

I'll never forget the intensity of the short, stocky man with his black, curly hair and piercing brown eyes.

"Bullshit," he bellowed. "We have obligations. That plane better damn well be fixed this afternoon."

Fire seemed to flare from his nostrils as he barked, "Now, get hold of some Cougar squadron down the field and get their mechanics down here now! Tell them to bring some nose gear door actuators and get that nose gear door fixed."

The entire Operations area was full of curious onlookers. No one said a word. We all edged toward the corners of the room. The forceful personality of the legendary Zeke Cormier dominated the entire building.

The lieutenant commander became a fast learner and barked "Yes, Sir!" He instantly grabbed the phone and started making calls.

Zeke and Nello ate lunch while mechanics feverishly worked on the crippled plane. The two left late that afternoon for their next airshow location. The Blue Angel tradition of making each commitment continues even today.

Several weeks later, I was jolted again with a reminder of the sacrifices paid by Navy pilots of that time as they flew their high-performance aircraft.

There were incidents aboard the carrier at sea, but there were tragedies happening right on the field.

At noon in early August, an F2H Banshee from one of the squadrons had been practicing strafing runs over the desert area east of San Diego and was returning to Miramar. The pilot reported in over the initial point, east of the field, and came streaking in at 1200 feet for the break doing about 250 knots. He was alone and ready to enter the break for a left-hand turn to circle down for a landing.

An instant before entering the break, the plane suddenly pushed over in a near-vertical dive toward the main runway. There wasn't enough time for onlookers on the ground to do anything except gasp. No one believed what they were seeing.

Within 2 seconds, it was all over. The twin-engine jet fighter impacted the runway in a ball of fire, scattering parts not much bigger than nuts and bolts in every direction.

Nothing remained of the pilot.

Fire crews had the flames out within minutes, but it took hours for the street sweepers to clean the last remnants from the runway. This was important to prevent fragments of the crash from being sucked into a jet engine of another aircraft and destroying it, or shredding the tires of other aircraft during landing or takeoffs.

As quickly as the runway was cleared, business returned to normal in both the launch and recovery of jets.

No one ever figured out the cause of the accident. Everything was analyzed, including the pilot's meals and hours of rest for the previous several days.

Somehow, a passion for flying and belief that mishaps always happen to the other guy—not you—manage to keep a military pilot or an eager young man motivated enough to press on in his quest to conquer the skies.

As a private pilot I flew during the afternoons and evenings at nearby Montgomery Field, and the lessons never stopped.

A buddy and I would usually rent a Cessna 150 and practice landings and takeoffs and occasionally take cross-country trips. We would do anything to be in the air.

One particularly damp, cold, dark, and cloudy night, Bob Cairncross and I crammed into the little cockpit of the rented 150 taildragger. We taxied out. There was no control tower.

The rotating white-and-green beacon on the south side of the field was blinking reassurance of the location of the airfield. Although we could see clouds and fog moving in from the west, for a couple young guys with their private pilots' licenses and about 60 flight hours under their belts, the fog should be no problem.

So much for youthful inexperience.

As Bob pushed the throttle forward, we gained speed. The tailwheel lifted and before long, we were up and flying—right into the fog bank that had moved in and was on the edge of the runway.

We were airborne, but we couldn't see anything!

I even looked straight down and could barely see any lights on the ground.

We had taken off on runway 26, which is 260 degrees on the compass, headed west toward the Pacific.

We climbed to about 300 feet to be sure that we were above any power lines.

We had to keep our speed up or crash. This was real life, and we couldn't believe what we had flown into!

What were we to do?

"Let me try it, Bob," I tried to say as calmly as I could.

I had about 15 flying hours' more experience, so I was the "expert." Gingerly, I banked 20 degrees to the left with the needle ball indicating an unbalanced turn. The ball was way off to the right side of the indicator. I was in a skid. Boy, was I nervous.

The airspeed was slow, a perfect situation to stall and spin.

We were close to being dead men, and we both knew it. But we weren't saying it to each other.

"Put more pressure on the right foot and bank more," I told my scared self. "Come on now, Dan, think."

"Bob, what's the opposite heading of 260?" I asked.

He thought a moment and said, "zero eight zero."

"Okay, I'm stopping the turn at zero eight zero. We'll just fly downwind for what feels like the right amount of time before we make a left turn back to the field. Hopefully by trying to fly a rectangular pattern we'll be lined up with the runway."

"Okay, watch your altitude."

"Keep looking for the runway lights," I demanded, "these red instrument lights don't glow bright enough. It sure makes it hard to read the gauges."

I tried to fly what felt like a normal rectangular landing pattern.

"Okay, I'm turning final."

I leveled out at 300 feet on a compass heading of 260 and hoped that the runway was ahead of us. Then I started a slow descent.

I looked over at Bob for a second, he was peering straight ahead. We hadn't been talking much.

"Look out for the runway," I shouted.

He already was.

Airspeed 80 miles an hour. Fast enough.

We dropped through 200 feet, then 150 feet, with nothing but white fog in our faces.

I kept the compass steady. One hundred feet.

We broke out.

Way over to the right were the runway lights.

My rectangular pattern hadn't been perfect—but it was good enough. I corrected sharply to the right, lined up with the runway, and dropped down for a bounce landing.

Both Bob and I were trying to act calm as we taxied in to park the aircraft. It had been a frightening experience. My legs were shaking so bad I wondered if I could walk.

In our first casual banter as we tied the airplane down, we both vowed we'd never take a chance with the weather like that again. We were lucky.

I gained instant respect for weather instruments and instrument-qualified pilots.

All the activities at Miramar and the weekly steak fries at the "O Club" had me full of aviation enthusiasm. I was glad that the end of summer had finally arrived. It was time to head for Pensacola and my chance to fly with the big boys.

CHAPTER FOUR

Flight Training

FINALLY, PENSACOLA—the cradle of naval aviation. I arrived from San Diego in my new 1956 blue-and-white Ford Fairlane—the classic car of the day and a college graduation present from my folks.

It was packed with all the worldly possessions I'd need for the next year and a half of training. It had been a long journey. Entering the Pensacola area, I struggled to watch the road while gawking out the window to catch a glimpse of any training planes in sight. There were none—it was a Saturday afternoon. I passed a sign indicating Saufley Field, drove by Corry Field, and then got directions to the Pensacola Naval Air Station, better known as "Mainside."

Initially the airfields meant nothing to me, but they were about to become the center of my life. What a beautiful area! I saw sprawling green trees, thick grass, lush vegetation, neatly trimmed bushes and shrubs, nineteenth-century colonial wooden homes with stately porches on estate-size lots.

On the station itself there were only old solid dark red brick buildings and aircraft hangars.

The U.S. Navy had been training pilots here since 1914. During World War I, 999 pilots were trained here, and while World War II was going on another 28,000 began training here to earn Gold.

Now it was my turn. I checked into the BOQ (bachelor officers' quarters) and found the nine other young ensigns from my NROTC class at the University of Missouri. We were now a part of the Aviation Officer Class 01-34-56. We'd gone through college together, and now were starting navy flight training together. But all of us ended up going in different directions by the time we got our wings.

A new student pilot arrives in Pensacola gung-ho, excited, and fired up with enthusiasm to climb into the cockpit and get started with navy flight training. He's on such a natural high just to have been selected. It seems there's nothing in this world that could discourage his zeal.

Church on Sundays was also important. I came to appreciate the Navy hymn called "Eternal Father." It is truly an inspirational song for a young

Naval officer with a dream to fly. A recording of the song written by John Dykes and produced by Eddison von Ottenfeld said it all.

Eternal Father

Anyone who has seen a pilot, lift the wings of his plane from the deck of an aircraft carrier, or watched a plane plunge into the sea, knows the deep plea of the words:

> Be with them always in the air
> In darkening storm or sunlight fair
> Oh hear us when we lift our prayer,
> For those in peril in the air.

Inspirational music of the Navy hymn "Eternal Father" best expresses the great, the quiet and powerful devotion of the American fighting man to God—his Eternal Father.

> Eternal Father, strong to save
> Whose arm doth bind the restless wave
> Who biddest the mighty ocean deep
> Its own appointed limits keep
> Oh hear us when we cry to thee
> For those in peril on the sea.

> Lord, guard and guide the men who fly
> Through the great spaces in the sky
> Be with them always in the air
> In darkening storm or sunlight fair
> Oh hear us when we lift our prayer
> For those in peril in the air.

It didn't take long for reality to set in. The first day of orientation we learned we would not get near an airplane for the next 6 weeks. First, there would be over 200 hours of ground school instruction plus grueling physical fitness training and water survival. Before flying we were immersed in engineering, aerology, navigation, and principles of flight.

The Navy wanted us to learn about engines, turbines, propellers, wings, and clouds before we got near any of them. It was logical, but like college all over again. Most of us expected just to jump in a cockpit and take off for the wild blue yonder.

The first day of class was another jolt. Our first class was speed reading. Speed reading! I thought I came here to fly.

"This will help you learn what you need to know faster. You're going to find this training more demanding than college," we were told.

By the time we acquired all the textbooks and training materials, we knew we had to be in good shape just to lug them around.

But that wasn't the end of it. The PT (physical training) was rugged and demanding. Two elements stand out in my mind—the step test and the obstacle course. The step test was simple enough, just stand in front of a 16-inch-high bench with a long row of other students and step up onto it, in unison. First, you put one foot on the bench, then stand up on it with both feet; you then step down backward with the first foot, then stand on the ground. You then repeat the cycle as fast as you can—for 5 minutes.

The instructors are shouting at you to go faster and chiding you if you even think about slowing down. Guys begin to drop like flies. The first minute or so is nothing, but after 3 minutes it is serious business. It was all designed to create competitive spirit, and if you didn't make the required 5 minutes, even with remedial help, there was no flying in your future.

The obstacle course was the ultimate torture weapon. It was set in a wooded area where you couldn't see around the winding, sandy track of barricades, rope climbs, and impediments. It was designed to test one's agility and endurance and virtually every muscle of your body.

If you didn't make it around in 3 minutes 42 seconds, you were history. The record was 2:25. There were dual lanes of old rubber tires to do a high-step running waddle through. Then you'd crawl on your belly in freshly watered-down sand under pipes about 2 feet above the ground. If you raised your head too high while squirming through, you ended up with a knot the size of a walnut. Then there was a solid wooden bulkhead 12 feet tall with a 2-inch rope to grasp and pull yourself up and over and then jump to the ground. Then there were high rails to balance on as you ran along too high in the air to want to fall, with instructors waiting to make you start over if you did. Then there were Jacob's ladders to climb and platforms to climb and jump.

By pitting us against one another, the competitive spirit took over and motivated us even more. You knew that your time would be posted for the entire class to see.

If I had suffered during my brief stint of football in college because it nearly killed me, I was glad to be small here. The big, heavier boys who struggled to finish in the maximum time allowed were singled out for special treatment and extra exercise by the enlisted instructors.

Water survival was special. As a Navy pilot you were expected to be a good swimmer. There were all kinds of swimming tests—the ultimate was the Dilbert Dunker. It was simply a crude metal-simulated airplane cockpit attached to a platform about 30 feet above a gigantic swimming pool. You climbed up a ladder, stepped into the cockpit, and fastened your seatbelts and shoulder harnesses with a simulated seat parachute strapped securely to your rear end.

When the instructor felt you were ready, he would release a latch, and this metal box with you inside would scream down a 45-degree rail, hit the pool water, flip upside down, and slowly sink toward the bottom. It was intended to simulate an aircraft water ditching. Your job was to keep from getting disoriented and to get yourself unstrapped, swim deeper into the pool and off to the side of the simulated aircraft, and then come to the surface. All of this was to be accomplished without drowning or needing the safety instructor. He was using scuba tanks at the bottom of the pool ready to come to your rescue.

This event required extreme self-discipline, not just to survive the Dunker, but to appear nonchalant and casual to your classmates while your stomach was churning and your nerves were jittery. You wanted to be sure to do it right and not screw up in front of everyone lining the pool watching and waiting for their turn. You also didn't want to look stupid to those who had gone through the test successfully before you.

We worked for 6 intensive weeks filled with classroom studies during the day, PT in the afternoon, and study hall in the evening. Some of the unbridled enthusiasm began to wane as scholastic demands and physical conditioning began to take their toll.

Some of the guys began to question why they selected such demanding training, especially when the only airplanes they had seen were the ones flying overhead, destined for some faraway landing field.

Aviation author Kit Lavell spelled out the reason for the tough training:

> During a visit to Pensacola I frequently heard the same questions among the students that I asked myself twenty years before. "I'm a college graduate. Why do I have to have some guy scream at me what to do?"
>
> A former squadron mate of mine was being catapulted off an aircraft carrier at night. While being launched into a black hole, expecting to go from zero to one hundred sixty miles per hour in three seconds, he was unaware that his right landing gear collapsed, preventing his A-4 from reaching flying speed. "Eject! Eject!" yelled the air boss.
>
> He had a fraction of a second to obey and live, or die. After ejecting, he was run over by the carrier, but his survival training allowed him to escape unhurt.
>
> You've got to learn how to obey, and do it instantly in the military.
>
> Why are we punished if we leave our locker combination set one digit off the class number, "zero six"? I once helped investigate a fatal aircraft accident in which the pilot flew into a mountain because he had inadvertently set his radio navigation equipment one digit off.
>
> "Skivvies (underwear) folded four by four inches? You've got to be kidding" some students complained.
>
> Naval Aviation is a matter of precision: formation flying with inches between aircraft, landing on a pitching deck where the normal eight- or twelve-foot clearance between the tailhook and the aircraft carrier ramp might be a foot or less.

Learning to be precise can be the difference between life and death.

Then after the fourth week the word spread like wildfire. The Blue Angels were going to put on a show right here at Forrest Sherman Field at the Naval Air Station. That 40-minute demonstration flight of ground-level acrobatics charged us all up so much, almost everyone went home with dreams of becoming one of the "Blues." The next day we had plenty of energy to attack our studies with renewed vigor.

Finally ground school was over and it was off to fly airplanes at Saufley Field—and more ground school. Our studies became airplane-specific. We were introduced to the system that would follow us throughout our careers. Before you could fly any airplane, you had to study and understand what made that particular airplane fly and how all the systems worked, with special emphasis on emergency procedures. If something went wrong, the Navy wanted us to know every technique available to get out of that jam.

The slogans posted on the walls and printed in emergency procedure manuals simply stated, "The best time to know emergency procedures and the worst time to study emergency procedures is in an emergency."

There was no getting into an airplane until you had memorized cold all the emergency procedures—even though you carried a quick reference flip-chart-type manual of condensed emergency procedures in your flight suit leg pocket. In a Navy aircraft there wasn't time to sit back and go through a manual to figure out what to do next. It had to be in your head. Flight simulators hadn't been put to use yet.

Then one of the most special days in a student's life—issuance of flight gear. We all bubbled with excitement as we stood in line and were issued our flight logs. The clerk made notations in the back pages as each of us was sized and given his very own helmet, tan poplin flight suit, gloves, kneeboard, survival knife, and sturdy high-top flight boots. Finally we were issued the treasured leather flight jacket with "U.S.N." punched in the lining along the zipper and our very own names stenciled on white cloth patches sewn on the jackets.

If you earned your wings, you'd get the white patch replaced with a brown leather patch with stamped gold wings, with your name and rank printed under the wings. This is the ultimate status symbol to every student.

One of the married guys who lived across the street from me was so excited that he slept in his flight suit the first night he received it. His wife teased him and joked about it at parties, where students would trade flying stories and talk with their hands gyrating wildly in the air.

It may have been a little unfair and insensitive of her to make fun of her husband's love of flying. Heck, I felt like sleeping in mine, too! Every guy there did. That flight suit represented a real milestone in our efforts to become naval aviators.

We were preached to about being healthy and the importance of a nourishing breakfast on a day of flying. Yet I noticed that most instructors lived on coffee—a cup of black coffee to steady their nerves—and sweet rolls.

Breakfast before flying lessons at Saufley was at 0600 hours. It was barely daylight. We'd get cleaned up, don our flight suits, and stumble into the cafeteria for bacon, eggs, toast, and all the trimmings to give us energy to stand up to the rigors of flight and verbal abuse from instructors when we made a mistake.

Before our first flight we went through a bailout trainer. We sat in the cockpit of a T-34B with the engine running full blast, then jumped out the left side of the airplane through the propwash into a net to simulate a bailout. We also had to learn where every dial, switch, and lever was located in the cockpit, and be able to point them out to an instructor while blindfolded.

The dream had arrived—my first flight. I understood the T-34B inside and out. I was ready. My assigned instructor was a wiry, short, and slender Lieutenant jg (junior grade) Jones with a crewcut and face full of freckles. He was quiet and had a relaxing and disarming warmth.

My first decision was whether to tell him that I had a private pilot license. I elected not to tell. There were a lot of stories being shared among students, and one of them was that the students who bragged about having a pilot's license got their "comeuppance" from the instructor on the very first flight.

The stories included one in which the seasoned instructors would take up the student who thought he was a hot shot and ring him out with gut-wrenching loops, barrel rolls, slow rolls, split Ss, and spins. The instructors did whatever it took to get the student sick and bring him back humble. The enlisted ground crew was part of the conspiracy. They made the student clean up the cockpit, too. I felt it might be best to appear that I was a fast learner.

Our instruction included 12 hops before the first solo. They all went well. On my 12th flight the check instructor had me make nine touch-and-go landings at one of the outlining grass fields. Then he announced, "I've had enough of this. I'm getting out. Let's see if you can do it by yourself."

Panic, excitement, and a little fear raced through my entire being. I thought I was good enough to handle it, but still had to prove it to him— and to myself. This was a bigger and more powerful plane than the one I had flown in college.

I landed and taxied to the edge of the field. He unstrapped his seatbelt and shoulder harness, then buckled them together and pulled the straps tight so that they couldn't get tangled in the controls. He slid the rear canopy back in place and locked it shut.

He walked forward on the wings, slapped me on the shoulder, and yelled something I couldn't understand because of the roar of the engine and propeller noise. But I answered it with a grin and a thumbs-up.

He climbed down from the wing and walked over to the group of other instructors huddled together watching their students risk their lives on their first solo flights. They appeared obviously worried and nervous about their students. Our instructors were probably not as concerned about the student's welfare as they were about their decision and judgment to trust the student to fly alone, and the effect that would have on their careers.

It was lonely peering back to see that seatbelt all knotted up with no one there. My three landings went smoothly: one more milestone in my effort to get that pair of Gold Wings with the fouled anchor in the center that had been worn by Navy pilots since 1917.

A couple of flights later, my new-found pride was humbled. It was a flight to learn advanced techniques. With an instructor in the backseat, I taxied to take off from Saufley Field. After the takeoff I pulled the nose up to the climb attitude. This was old-hand stuff by now to a student Navy pilot, who was qualified to solo. But this plane wasn't responding correctly. I wasn't getting enough airspeed to climb.

I looked out the canopy to spot a vacant field in case I had to make an emergency landing. I rechecked the throttle. It was full forward; I practically bent it, pushing it against the stop. I scanned the engine instruments: oil pressure, oil temperature, cylinder head temperature—all normal. I mentally reviewed my emergency procedures.

There was total silence from the instructor in the rear.

Did he realize that we were in trouble? Finally I had to share with him the fact that we had a problem. I punched the intercom button on the throttle, "Sir, the plane's not climbing. I think we have a problem, but I can't figure out what it is."

"Is that right?" he drawled.

He didn't sound too frightened.

I scanned the instruments again. Everything looked okay, but we were not going anywhere and were about to stall out at this angle of climb.

There was a long pause, and then he suggested: "Why don't you try raising your landing gear?"

"Aw, what a dummy," I thought. With all that drag, no wonder I couldn't get the air speed up so that we could climb.

He laughed, and the incident kept me humble for weeks.

Throughout the training, while in the cockpit, there was constant emphasis on keeping your head on a swivel to spot and avoid any aircraft in the congested skies around the Pensacola area.

Several days after my humbling incident, the dangers of flying were highlighted again. One of the guys in the class behind mine was in the landing pattern and lost sight of the traffic ahead of and below him. He failed to tell his instructor, or to take action to spot the other plane, but stayed in the pattern, continuing his final approach.

Then Charlie Patterson heard a crunch of metal as his prop chewed off the tail of the T-34 in front of and just below him. The whirling propeller clipped the empennage just behind the surprised instructor's seat as both planes were letting down for the landing. Both planes crashed simultaneously to the ground; miraculously, no one was killed.

The student was allowed to stay in the program with a warning.

After completing primary flight training, we were off to Corry Field for transition to the more powerful SNJ.

Mine was to be the last class to use this vintage tailwheel trainer built in the 1940s. It had been used to train thousands of pilots before me, and was being replaced by the more modern T-28. But my class had to conquer this aerodynamic machine that could require full right rudder for directional control at slow speeds to prevent torque roll when power was added. The seat for the instructor in the rear had him sitting so low that he was virtually blind to what his student was doing. This kept most instructors' adrenaline flowing.

This monster plane taught me more about flying than any plane I ever flew. It was an honest airplane, but demanded the pilot's knowledge of flying.

If the controls were handled correctly, the plane responded correctly. But if the student was timid or indecisive with the controls, the SNJ could give him an incredible ride.

After we got comfortable in the SNJ, we began a syllabus in acrobatics. We learned the standard maneuvers of barrel rolls, slow rolls, aileron rolls, loops, half Cuban 8s, Immelmann's, split Ss, and spins—all combat maneuvers. I really loved this stage of my training.

My instructor was Marine First Lieutenant Goodwin. He'd let me try anything the SNJ could do. Although we were limited to a normal one-and-a-half-turn spin, he'd let me do three turns. We even attempted a five-turn spin. The plane really got cranked up, and I wondered for a few seconds if we'd ever get the rotating to stop, but we did.

We normally entered acrobatic maneuvers between 8000 and 10,000 feet. Standard instruction was to bail out at 5000 feet if we weren't under control with any maneuver.

The most scary part was on the first flight. Goodwin explained he flew the AD *Skyraider* (later known as the A-1) in a Marine squadron before coming into the training command. One of his buddies bought the farm because he didn't understand how a high-torque engine worked. They were doing low steep turns a couple hundred feet above the ground to keep track of some practice targets. His wingman pulled a steep 90-degree left turn too tight and stalled the aircraft. He kept the stick back and added more power. The plane flipped upside down and augered in, killing him.

Had he been in a right-hand turn, the stall would have rolled the plane upright making, an easy recovery from the stall.

Lieutenant Goodwin proceeded to demonstrate both maneuvers to me. I was apprehensive, but at 6000 feet we had plenty of altitude to experiment with. He taught me a valuable lesson. Confidence comes from knowing your aircraft and everything that it is capable of doing.

None of the planes I was to fly was an experimental aircraft. The limits had been tested, and the envelopes of the plane were known. I just had to learn how far I could go and feel comfortable with what others had pioneered. I got so that I felt I could fly the SNJ upside down better than anyone. Goodwin loved my enthusiasm and hunger for acrobatics and went to extra efforts to show me more than was required. By the time I was ready for my check ride with a different instructor, to prove that I had mastered this segment of training, I knew I was ready for the Blue Angels.

My weeks of practice were coming to a head. In a quiet, cocky way I was ready to wring out my check-ride instructor. That fateful day I managed to con the scheduling enlisted men to assign me my favorite and best-handling SNJ for the last great acrobatic ride. I had preflighted the plane with Lieutenant Judge quietly strapped in the backseat.

Strange, that is the first time a new instructor did not watch me preflight the aircraft, I thought.

I got in the aircraft, started it, and taxied out with just a few grunts from the backseat. We took off for the acrobatic area. I climbed toward 10,000 feet so that we could do the entire series of maneuvers. I had put them in a series and was going to show him my best stuff. All was quiet in the backseat. Then as we passed through 5000 feet, Lt. Judge instructed me to head for Mobile, Alabama, about 30 minutes away.

"But I'm supposed to show you my acrobatics," I exclaimed.

"I've got a hangover and can't handle any Gs or inverted flight today," he announced. "And your instructor told me you know what you're doing so we'll just do a little cross-country to kill some time and write off your check ride."

I've been robbed, I thought. What a disappointment. I had primed so for that particular hop and was being denied the chance to show my skill. I vowed to myself that I'd never do that to anyone. It's unfair not to be tested when you expect to be.

Night flying and formation flying were also a part of the SNJ training. First was formation flying. Those of us who thought we were pretty good got humbled again. This was to be a continuous experience in flying. There was always something new to learn about flying your aircraft.

Formation flying was a real challenge, especially CV or carrier breaks and rendezvous. The maneuver starts by meeting 6000 feet over some point, with an instructor in your plane the first time, and later under the watchful eye of an instructor in a chase plane.

To practice rendezvous, the idea was to turn away from your wingman, try to fly in a racetrack elliptical pattern, and rejoin him in close formation using as little power adjustment as possible.

At first we were all over the sky, jamming the throttle to the block or jerking it back to idle to avoid a midair (collision). Formation flying taught us more about being delicate on the controls than anything to date. We started out with two-plane sections then graduated to four-plane divisions. The poor guy flying number four in the echelon would get sucked badly out of position if the leader wasn't smooth and gentle on the controls or the throttle. We could sometimes get so far behind that it took agonizing minutes at full throttle to catch up. All of us would get chewed out, generally with some sarcastic comment by the instructor.

By the time we finished the 13 hops of joining up and rendezvousing on other SNJs and breaking away, plus a variety of turns, we were competent.

Night flying came next.

About 30 of us would be launched at dusk and orbit the field at 1500 feet for half an hour or so until it was pitch black. All we could see were the lights of houses and cars below.

Then we descended to a particular pattern and practiced landings in the dark with no landing lights. We lined up on the black patch of asphalt using runway lights as a reference to touch down, used some power, and climbed back into the black sky. There are not a lot of illuminating lights on ships in the fleet, so it was practice for the real thing—as dark as possible.

Keeping track of the wingtip and taillights of the airplane ahead was important in avoiding a midair.

My classmate, who had sliced through the T-34, was orbiting with us one evening and had engine failure. He made a virtually blind nighttime emergency landing in an unlit dirt area of the field that was just a black mass. He landed gear up on the belly of the plane. The only damage to the plane was a few antennas torn off from the belly. The prop had stopped parallel to the wing, so it wasn't damaged. He had failed to radio what was going on so that the rest of us could get out of the way to avoid a midair.

The accident board determined that the cause of the accident was the student's failure to follow emergency procedure. He forgot to periodically switch fuel tanks and ran the fuel tank dry in one wing. Then he forgot the emergency procedures drilled into his head to switch fuel tanks the first instant of engine failure. The other wing tank was full of fuel. He must have had a guardian angel. To my amazement, they let him stay in the program until he crashed an S2F when he advanced to multiengine training in the Corpus Christi area. The Navy finally decided he wasn't a good risk.

Rain in Pensacola is common, and when it rained, all flight training was canceled. But students didn't get to go home on the off chance the weather would clear up. So we were kept sitting around the hangar. We'd study emergency procedures until we knew them forward and backward. We were very bored.

A lot of us spent dozens of dull hours with headsets clamped over our ears in the code room. Learning Morse Code by ear and flashing lights was required. We had to write down 20 words, or actually 20 blocks of five random letters, a minute if by ear and eight by flashing light. The sheets were graded. It was tough, but it was necessary to identify navigation aids in those days since the A and N signals were the standard device and VORs (very-high-frequency omnidirectional radio ranges) were just coming into existence.

The other way to kill time waiting for the clouds to clear was playing card games—especially hearts. We all took great competitive delight in sticking our buddies with the queen.

Plenty of stories were also exchanged about events at a local beer joint called Trader Jon's.

It was a dingy, dusty, dark, high-ceiling room like a barn with a giant U-shaped bar in the center. All types of flying memorabilia hung from a false ceiling. There were model airplanes, helmets, scarves, cut ties, and dozens of other types of souvenirs—each with some story from their owner before it became part of this poor man's museum.

On one wall were autographed photos of squadrons, planes, heroes, and some destined to become heroes. Many of the picture frames hung at a crooked angle. There were also dozens of plaques with squadron patches—from both the training command and the fleet—plastered all over the walls. The place reeked with aviation history and stories while students and instructors soaked up the suds and girlfriends listened in awe in the process of trying to hook a highly eligible bachelor.

Some of the knicknacks looked like they had been there since the Wright Brothers started flying. There must not be a naval or marine aviator who has not been to Trader Jon's. It still exists.

After we had learned how to fly acrobatic, formation, and night flying, next was basic instruments. Now we reversed seats with the instructor in the SNJ and were confined to the backseat with a white hood pulled over us. It was like a cloudy skylight, so all we could see was the instrument panel. No peeking allowed, besides it was designed so we couldn't cheat.

Sitting in the backseat was like having your head in a long enclosed cardboard box with some instruments at one end and your eyeballs at the other trying to figure out what the instruments were saying. The instructor started the training part of the flight with unusual attitudes. First he'd have me bend over, stick my head between my knees, and close my eyes while he controlled the airplane and did wingovers, dives, and near-stalling climbs. Then with the airplane virtually out of control he would yell at me, "You've got it," and it was my job to figure out what was happening to the airplane by only watching the instruments and get it back under control.

It was dull work for the instructors to sit there wobbling all over the sky with a student in the backseat learning to develop an instrument scan.

Lieutenant jg Clausen was my instructor. The idea was to get the student skilled enough to fly a smooth Charlie pattern. This is a timed series of circling climbs, glides, and turns designed in such a way that you flew a giant square in the sky and ended up where you started. It took total concentration to keep the gyro at the proper attitude, and correct turn while working the throttle and watching the manifold pressure to keep the airspeed exact for the climb and glide. All the while you must watch the second hand on the clock to be sure your timing was right.

Everyone hated this part of the training including the instructor, who thought it was dull, boring flying. The SNJ was noisy even with the canopy closed and was like a greenhouse in the summer, especially with the hood over your head.

After a while I got pretty good at this basic instrument training. Not great, but good. Clausen was excellent at creating confusion for the poor student in the backseat. My instrument scan and concentration were under control one day—then suddenly there was a loud noise. The forward canopy slid back on the tracks and hit the stops with a loud bang. The wind blew in, scattering dirt and dust everywhere and flapping the checklist, charts, and papers on my kneeboard. The cloth hood ruffled and flapped.

With the sudden blast of wind I wondered if there was something wrong and whether the instructor was going to bail out. I couldn't see a thing with the hood over my head. My instrument scan went to pieces. No telling what direction the plane was going. I was rubbing debris out of my eyes. The intercom came alive in my ears with the demanding voice of Clausen barking, "Come on numb nuts. What the hell is going on back there?"

I thought to myself: "He just opened the canopy totally for no reason at all at 6000 feet in an effort to disrupt my flying. My concentration is gone. I've got vertigo so bad I can't read the instruments. We're lucky we're not in some graveyard spiral. The plane's out of control. What the hell does he think is going on back here? We're out of control. That's what's going on back here."

Eventually I learned to fly basic instruments in spite of Clausen's nasty tricks. I felt like telling him I was going to fly inverted the next time he pulled some hairbrained stunt on me, then push the stick forward and try to dump him out of his open cockpit.

This finished my basic flight training. Now I was supposed to go to advance training and carrier landings. Right at this time the Navy changed the entire flight training program.

First, they started by insisting that students, in a couple of classes behind mine, sign up for 5 years of active duty after they received their wings instead of the 2 years that was the original deal.

The Navy had calculated that they were spending too many dollars training pilots and not getting their money's worth with such a short active-duty commitment. They were right.

Second, carrier landing training was no longer available to everyone. Unless you were going to a squadron operating-carrier-based aircraft, the traditional event was stricken from your curriculum.

I had just decided to select helicopters. With all the turmoil changing the syllabus, the Navy didn't know what to do with me and a dozen or so other students. It wasn't worth spending the extra couple of months to fight the system in Pensacola to get the six carrier landings. We were already at Barin, the field where we were supposed to prepare for carrier landings. While we were waiting, the word came down to give us some SNJs and let us go flying to keep up our proficiency.

That was nearly a big mistake.

On the first couple of hops we made, each guy did whatever he wanted, mostly acrobatics. Then we started talking. We all figured we were pretty good at acrobatic and formation flying. If the Blue Angels could do it, why couldn't we?

So, with no briefing, on the next hop we decided to rendezvous 10 miles north of the field above the intersection of some highways at 7000 feet. We rendezvoused in echelon and did a few turns together. Nothing to it.

Then we decided to switch to a diamond formation. We were not as tight as the Blues and their jets. After all, we had to be careful not to have some stupid accident by chewing a couple of planes' tail feathers with the props. A few more gentle turns in the diamond, then steep turns. We all hung together pretty well. Next came wingovers. We were doing great.

One thing led to another.

I don't know who had the bright idea, but next we tried a loop. Down, down, down to gain airspeed. There were some cumulus clouds in the sky and the air was unstable. We bobbled around as we picked up speed— finally 185 knots. I was back on the stick, eyes on the leader, watching the horizon out of the corner of my eye. Backpressure, wings level, up, up, and over the top, we were slowed to about 80 knots. Then we started down. About that moment a voice burst over the radio.

"What the hell are those guys doing over there in that formation?"

Obviously an instructor had spotted our antics and was radioing to another instructor. Did he know who we were? I was convinced he was going to find out. We were still in formation heading straight down on the recovery from the loop maneuver. I got real nervous. Visions of the end of my career in naval aviation flashed into my mind.

Why did I ever do this, anyway? I thought.

My stomach instantly knotted.

Then I figured since there's one instructor and four of us, he can't catch all of us if we split.

We all must have thought the same thing simultaneously.

Heading down in the vertical, I twisted and turned my SNJ away from the formation; so did the other guys. On the way down on the backside of the loop my airspeed was already increasing. Now with my concentration on the sudden turn to get away from the formation, I forgot about airspeed. Then I looked at the gauge: 180 knots.

I was still about 45 degrees nose down and going through 5000 feet— the base altitude for acrobatics and the altitude at which we were supposed to bail out if we didn't have our plane under control.

I pulled back on the stick. The Gs built up. I pulled more than I'd ever had up to this point in my training.

I finally got leveled off at 4000 feet and 201 knots. The red line was 206 knots.

Phew! I just sat there and let the excessive airspeed bleed off while I plotted what to do next. I was too scared to try acrobatics or anything else.

My heart was pounding so loud I thought I could hear it above the engine noise.

I just wanted to get back to Barin undetected. Should I go straight back, or what?

I dropped down to 2000 feet and headed east. Since we were north of the field I decided to make a big circular arc flying about 10 miles from the field and approach Barin from the south. About 40 minutes later I called the tower for a landing clearance and received it.

Once I parked the aircraft and shut down the engine, I walked to the parachute loft to return my chute. A couple of the other guys in our formation were loitering around the classroom area, casually checking out for the day. So did I, as quickly as I could. No one wanted to be around should someone start asking questions. We hardly looked at each other, much less talked. I couldn't get out of there fast enough that afternoon.

The next day we all flew alone.

No one ever said anything to us. Were some instructors communicating to each other, or were they talking about us? We never knew.

But one thing I did do the rest of my time in the training command was to follow all the rules. Getting those Navy Wings of Gold meant too much for me to blow it on something stupid.

Now the Navy decided to give us more instrument time in the twin-engine SNB. They teamed two students with an instructor. We'd fly for a three-hour hop with each student switching from pilot to passenger after 1.5 hours. My mate and I got Marine First Lieutenant Ryan for our instructor, and we were his first students. He was a "plowback"—a student who'd graduated and was now selected to be an instructor rather than being assigned to a squadron in the fleet. Not a career enhancing event and he wasn't happy about it.

Ryan was a nervous wreck trying to show us how to fly instruments. He was yelling and shouting throughout the entire first flight. He was not a skilled teacher, and both of us turned in a miserable performance of instrument flying. The only thing that saved me from my first "down" or unsatisfactory flight was that Ryan knew he did a lousy job too and was lenient in his grading.

The next day we declared an unspoken truce between students and instructor, but the flight didn't go much better. My partner and I had visions of our entire careers as Navy pilots evaporating. Sooner or later Ryan had to give us a down.

Pairing and scheduling students with instructors was done by a third-class petty officer who tended a giant blackboard in a corner of a hangar. My partner came up with the bright idea of doing a little bribing of the enlisted-man scheduler. I didn't believe in such a thing, but this time I sure wasn't going to oppose it. My partner gave him a fifth of Scotch and the next day we were scheduled with a different instructor, to the relief of my partner, me, and Ryan. Everything went great with the new instructor.

The routine was to fly a 3.0-hour hop. One student would fly the first half and then climb back in the cabin and the other student would take over. One and a half hours on instruments is about as long as any student can concentrate on learning navigation by instruments and radio signals. The nonflying student was supposed to sit on one of the small seats on the left side of the plane, behind the flying student, and look out for other planes in the area, so we would avoid a midair. The flying student pilot was absorbed in his instrument technique. But the general practice was for the nonflying student to lie down on the floor of the aircraft and take a nap.

The SNB had five fuel tanks. This meant you had to keep a constant eye on the fuel gauges and switch tanks before the plane ran out of gas. It was difficult to remember the fuel gauges when you were concentrating so hard on the instruments and flying properly—especially with an intimidating instructor next to you.

One day I was stretched out on the floor sound asleep with the monotonous droning of the engines helping me sleep. Suddenly there was silence and no vibration.

We had run out of fuel. Both engines had quit.

From a deep sleep I bolted up, instantly figured out the problem, automatically shot forward, switched fuel tanks, grabbed the wobble pump handle on the floor next to the student pilot seat, and furiously started pumping fuel back into the engines.

Finally the engines got fuel into the carburetor and ignited and the windmilling props started churning under power again.

All this happened so fast that the student pilot didn't even have time to react and fix the problem himself.

The instructor was furious with me. He had spent the last 20 minutes keeping track of the fuel gauges to see if my partner was going to run out of gas while he was busy concentrating on instrument flying. Then to run out of fuel with plenty of altitude and not have the chance to see how his student responded to this semiemergency situation made the instructor fit to be tied. I was just protecting my own hide.

Eventually we were required to qualify in the SNB. Up to this time we had used it only as a platform to learn instruments.

Learning to fly a twin-engine aircraft wasn't all that hard, because the engines were very similar to those of the SNJ. The student would always sit in the left pilot-in-command seat with the instructor in the right seat. We learned to fly single-engine emergencies at altitude and then down in the landing pattern.

Many times just when we were airborne, the instructor would reach over unexpectedly and pull back on one of the throttles to simulate an engine failure. One engine would idle. It was our job to push the other engine forward and make corrections on the controls and trim to keep the airplane flying. We had to follow the correct procedures, and return to the field and land safely, all on one engine. However, the other engine was only at idle and was ready if necessary to have both engines working or in case the other engine should actually fail in flight.

One of the best instructors I had was Lieutenant jg Trimball. He loved to practice emergency procedures. After a few flights with him I didn't think there was an emergency I couldn't handle.

One day at an outlying field, we were the only aircraft in the pattern. The area was deserted and no other students or planes were in sight. The other student who was supposed to fly with me failed to show, so Trimball and I launched on the scheduled training flight.

We flew instruments on the way to the field and he called out a simulated ground-controlled approach (GCA). We shot a touch-and-go landing, and immediately on the climbout Trimball pulled back the throttle on the right engine. I went through all the usual emergency procedures, turned into the good engine, and started flying downwind for a landing—one engine powering the airplane, the other back at idle with the propeller windmilling.

"You think you can land this plane single-engine?" he asked above the noise.

"Yes, Sir," I said, assuming we were going around the pattern with one engine in idle—as I had done so many times before.

With that he reached over, pulled back on the right mixture lever to idle cutoff, and told me to hit the feather button. Quickly and instinctively I did. The engine totally quit, and the prop windmilled until it stopped. There was no movement by the prop. It just stood straight out there. We truly were single-engine. All my self-discipline and concentration was focused on following the procedures I had learned to get the aircraft safely back on the ground in this situation.

However, it dawned on me that this was real. This was a self-inflicted emergency. Trimball had real guts. He just sat there cool as a cucumber. He didn't show any emotion or say anything. He just watched me fly. I continued downwind, turned base, then final. All the while I prayed my good engine wouldn't quit. He still just sat there. No words of advice. No comments, nothing. I made an uneventful full-stop landing.

Soon as we came to a stop, he reached over, hit the starter, cranked up the dead engine, and said, in his usual casual nonchalant manner, "You were right, you can make a single-engine landing. I think you have the hang of this. Let's go do some navigating."

Both the fright and confidence that came from that one maneuver never left me. The fright instilled in me the need to always be prepared for any emergency. And the confidence was that if I was prepared, I could handle it.

Cross-country trips were part of the training. One of them was a trip to Floyd Bennett Field just outside New York City. It was me and my instructor. We made it with a couple of refueling stops. Those allowed me to get in some real instrumental flight rule (IFR) approaches. Then I learned one of the tricks of the trade. The instructor wanted to stay in

New York for several days. His way to do that was to keep grounding our aircraft for any mechanical reason he could dream up. I was enthusiastic to fly, but became a prisoner of his wishes for a good time in the big city.

The change in the flight syllabus and excellent weather helped me finish faster than the normal 18 months with a total of 214 pilot hours (200 were required). I completed training in almost 10 months to the day since I first climbed into the cockpit as a student.

Then the big day came, September 11, 1957. I was all decked out in my new khaki uniform with fresh pressed creases for the ceremony as my wife pinned on my Wings of Gold with the Navy League of the United States embossed on the back. I was now excited Navy pilot number T-6280. The dream had come true.

It turned out to be just the beginning. It seemed like I was to learn something new on nearly every flight. You just never got to the point where you could take anything for granted.

Instead of going to the fleet as a pilot, my orders had me for an additional 8 weeks training at Ellyson Field to qualify as a helicopter pilot. The instructors at Ellyson joked that once you qualified in helos you were an "unrestricted" Navy pilot, meaning you could fly both fixed-wing and rotary-wing aircraft.

Ellyson Field was 15 miles north of Pensacola. The helicopter training field was named after the Navy's first aviator.

Once again, it was back to ground school. About the only relationship between helicopters and airplanes is they both move through the air. Some of the control movements are similar, but it's really like learning to fly all over again. The helicopter stays in the air because of a whirling blade overhead. Some kind of tail rotor determines the direction in which you want to head.

It all sounds simple, but the helicopter is a totally unstable flying machine. I was to learn many fundamentals of helicopter flight.

Helicopters at this stage of their existence had no trim tabs, no boosts, no assist, no computers, no navigation helps. Helicopters needed two hands and two feet staying in constant motion pushing, pulling, and lifting the controls with literally hundreds of correcting movements.

It didn't take me long to discover that there was never a chance to hold just the controls steady. I was always correcting and countercorrecting to keep the mixmaster of a machine heading in the direction I wanted it to go—or as a beginning student—in the direction I wished it to go. Helos at this time used gasoline-powered engines and had a tendency to catch fire.

There was never a chance to let go and rest. Your hands and feet were married to those controls. You let go, and you're out of control.

In a helicopter the engine needs to operate at near-maximum power throughout the flight to keep the rotor blades spinning fast enough to keep the helo in the air. This makes it easy to overspeed or wear out the engine fast.

When reciprocating engines continually run at top speed, they don't last long. That proved to be true with the HUP.

Racecars have many engine failures because they always run at maximum rpm.

I was taught that you cannot be careless or abusive with the controls. If you are, you could have a nice quiet ride with the rotor blades thrashing and the wind rushing past the helo as you silently plummet to the ground in autorotation with a blown engine.

There were other key lessons to learn. If you pulled up on the collective with your left hand, all blades of the rotor system would take an equal bite of the air to lift the helo up. Because of the higher angle or bite of air with the rotor blades, more power is required to keep them rotating, so you must smoothly add just a little more power to the motorcycle handle grip on the collective to keep the blades from stalling. Too much power, and you'll overspeed the engine; not enough, and the blades slow down and eventually collapse or cone upward or stop rotating. The result is, you fall out of the sky. It's a delicate balance and all manually controlled by the pilot.

Roughness on the cyclic makes the helo dance all over the sky. Near the ground in a hover, the roughness can be dangerous.

The performance of the helicopter is also affected by the air through which you fly. It's called "density altitude." If the air is hot and humid, it has a double-whammy effect on a helo. It decreases the efficiency of the engine performance just as in an airplane, and it also hampers the rotor system from effectively biting into the air to provide lift. This can be compounded by the wind or lack of it and altitude.

Some wind, on landings and takeoffs, is helpful to provide lift since the rotor system is simply a rotating airfoil. However, too much wind can have an effect on the blades, causing them to slam into the side of the helicopter. I saw results of this type of accident from slow rotor speeds on engagement or shutdown many times. Fortunately it never happened to me.

Once you understand these basics, flying a helicopter is just a matter of going out and doing it, if you can.

The first lesson to learn in the air is to relax and not be tense at the controls. What a joke. This is easier said than done since you're so busy and so scared of the contraption that you find yourself praying you'll merely survive to the end of the flight.

During my first flight the instructor tried to explain the different types of vibrations, which include high- and low-frequency vibrations, out-of-track rotor systems, and more. All are precursors of problems. Seated side by side, he'd talk to me and try to size up my reaction as he explained what was happening. After he'd ask me for the third time, to keep from sounding like an idiot, I'd nod my head affirmative, punch the intercom button, and acknowledge "roger" over the headset. The lucky thing was that my helmet sunvisor was down and he couldn't see the bewilderment in my eyes trying to figure out what he was talking about.

Strutting around after earning my Navy Wings of Gold didn't last long because I was being humbled just when I thought I was pretty hot stuff. This new way of flying and the technique is a lot different from that of an airplane.

You learned to make throttle adjustments with a slow squeeze.

When I lifted up on the collective, torque caused the nose to turn to the right, so I'd need to add left rudder. If I overcontrolled the rotor, it changed the heading, which affected the power setting, which meant that I needed to crank on more power or let off on the power.

The power is indicated by the pair of rpm needles—one for the engine and one for the rotor system. When they are married, one overlaps the other, indicating that the engine and rotor blades are working satisfactorily.

One needle is short and indicates rotor rpm on the inside of the gauge. The long needle points to the engine rpm. If the needles are not overlapping during flight, there is a serious problem. This is the most watched gauge on the instrument panel.

If you raise the collective too high, you also need to adjust the power setting. It's all done manually.

My introductory flight was in the gasoline-engined HTL-6—a metal-frame bird seen in the "M*A*S*H" TV show with a small Plexiglas nose where the pilot and copilot sit. My instructor was Lieutenant Robinson.

After avoiding airplane stalls that could progress into a spin and death at low altitude, Lieutenant Robinson could see the apprehension in my eyes when we flew up to 500 feet, and he said, "Watch this." He pulled the cyclic back, and the airspeed indicator went from 60 knots to 40, to 30, to 20, to 10, and then 0.

Zero—I was grabbing anything I could hold onto—the frame of the cockpit, the edge of the canopy, anything except the controls. I didn't want to touch them so that the instructor could use them to get us out of the spin I assumed we were sure to enter. Besides, I didn't know what to do with the controls. Nothing happened—we just hovered at 500 feet.

After my initial fear subsided, I thought this could really be fun.

"Here, you try it," he told me.

We gyrated around a lot and I was rough on the controls, but he had plenty of patience.

After my first flight, I got a new instructor, Commander Luke. He was older and soft-spoken. He was also gentle and frail, but unflappable.

"Now we're going to learn how to control the helicopter down low near the ground. This is a machine you've got to be very precise with," he stressed.

"Hovering is supposed to be near-motionless flight," he explained.

But it took me awhile to keep from yoyoing all over the practice field as I learned to coordinate control movements. Air taxiing is basically hovering while moving the helo about 5 feet above the ground without bouncing up and down or hitting the ground.

There were big squares for guides, and I was dancing all over the place. Suddenly I learned it wasn't small movements on the collective, but just small pressures on the controls, to maintain a level attitude and use of the rotor to maintain the desired heading.

The real effort was to get rid of tension, which caused late reaction. This always caused a new student to be behind the helicopter movement. I was no exception. I also had to learn to allow for a lag in cyclic and collective controls, which would lead to overcontrol. But in the Bell HTL, the simplest of all helicopters to fly, it did not take long to learn hovering, air taxiing, and vertical landings. Next was the transition to forward flight from the hover. It doesn't sound like much, but the first couple of times were tricky.

"You smoothly and slowly ease the cyclic stick forward to gain airspeed and follow immediately by adding collective pitch with the throttle," explained Commander Luke. "This prevents settling as we fly off the ground cushion created by the down draft of compressed air created by the main rotor blade that helps support the helo."

"Someday this could be a tricky maneuver when you're at max power," he reminded. And it was during later rescues.

He showed me all the basic airwork maneuvers necessary for vertical flight and how to handle the control movements for forward flight slow cruise, fast cruise, climb, glides, approaches, steep approaches, and side-ward and rearward flights. Then I attempted to put all the basics to work together doing constant headings, squares, parallel heading squares, perpendicular heading squares, figure eights, and turns on a spot.

"You've pretty well mastered everything today," Commander Luke told me one morning.

"Today we are going to try autorotations."

"Autorotations?" I thought, as my stomach muscles tightened. I knew to anticipate this day. Every student had studied what happens when the engine quits on a helicopter. If the controls are handled properly, the rotor blades act as a giant windmill as the helo heads for the ground, whether the surface is level, hilly, rocky, or water. But as the autogyro proved years before helicopters were produced, there is a surprising amount of lift produced from the spinning blades. You needed to time it just right when to exchange the energy of blade rotation for lift as you head toward the earth's surface. You're going down. How you handle the controls determines how you and the helicopter will survive the engine failure.

"We'll try our first autorotation going straight ahead," he briefed.

"We'll just lower the collective pitch rapidly all the way down and hold it there, simultaneously reducing engine rpm from 3100 to idle at 2500, which will split the needles and separate the engine from driving the rotor blades. Then we'll hold the collective full down to keep the rotor rpm fast enough, so the rotor blades don't stall as we head for the square in the field. Considerable right rudder will be necessary to prevent skidding resulting from the change in engine torque.

"Keep the cyclic forward, we'll maintain 45 knots forward speed," he continued.

"As you pass through 200 feet, be sure you're headed into the wind and start to pull back on the cyclic to slow down airspeed. When you get to 10 feet, smoothly increase the collective pitch and add power. Marry the needles so you stop the rate of descent by 5 feet, add more power, and wave off. We don't want to touch the ground on the first time—we just want to practice the technique. "How does he expect me to remember all this," I wondered.

"We need to practice autorotations because it's the only way down. No parachutes if the engine quits. Even if you did have a chute, you would never get out without it being sliced up by the rotor blades. So you've got to learn how to do this real good—there is no second chance."

With that, we lifted off, climbed out, and flew to a practice field.

"All right," he told me, "Let's try our first autorotation. Get lined up and lower your collective."

For a moment I wondered who'd be the most nervous—him or me. Then I looked at the windsock on the ground 500 feet below and made sure we were headed into the wind. Taking a deep breath, I firmly pushed the collective all the way down and held it there. Suddenly the airspeed slowed.

"Forward on the cyclic," Commander Luke quickly ordered. I got the airspeed back up somewhere around 45 knots. The ground was approaching fast. I mechanically put in the right rudder. We were still drifting left; it needed more right rudder. My right leg pushed forward more.

Two hundred feet already, the engine rpm was down to about 2000 rpm—too slow. When I added collective at the bottom, if it was jerky, the engine rpm could damage the clutch when it connected with the rotor. It would also drag down the rotor rpms and keep us from getting airborne. I tried to gently add more engine rpms so that it would be easier to marry the engine and rotor needles at 10 feet.

Things were happening so fast I felt that I was just along for the ride. Commander Luke was following along with his hands and feet on his set of controls. I really wasn't sure how much of the flying he was doing and what I was doing, but I was thankful he was there. After we passed 200 feet from the ground, there was time for only one more glance at the instrument panel at about 100 feet. Then it was all eyes outside the bubble canopy to estimate when we were about 10 feet off the ground so I could add collective before we hit the ground. When we got there, I pulled up on the collective and added power and the helo jolted as the engine took over and we climbed out for a second try.

"Not bad," Commander Luke commented.

I really wasn't sure what I had done, but I wasn't about to confess that to him.

"Now let's see you be a little more smooth and think ahead on the next one."

I gradually improved and finally learned enough so that he'd trust me to do a full autorotation to a sliding touchdown on the skids.

Then we did 90- and 180-degree turns into the wind autorotations. Getting the hang of autorotations was truly a confidence builder. It taught me how to finesse the controls and believe that I could safely operate the helo in any condition.

After nine flights and 8.3 hours, Commander Luke declared me safe for solo. The solo flight was the ultimate experience.

On a fixed-wing flight you always figure if the engine quits, you would just glide the plane to a landing in some field. You would have plenty of metal surrounding you for protection, so you could just climb out; or your parachute would let you safely down, hopefully without too much damage to your body.

But there is something about being a helicopter pilot. You're always sitting on the edge of your seat wondering not just if the engine is going to fail, but when it's going to fail and if you will be within the flight envelope to ensure a safe landing.

We were instructed the deadman's curve was between 200 and 500 feet. That's the dangerous area to fly in and vitally important to keep airspeed high enough to ensure a better chance for success in any autorotation. The problem is, in case of engine failure, that the rotor system needs time to recover the rpm that decays during the brief lull from the time the engine quits until the pilot gets his collective full down. This reduces the bite the blades take in the air so that they can windmill at optimum speed.

I spent 24 flight hours perfecting and smoothing out my helicopter flying technique in the HTL, including cross-wind and backward takeoffs, engine failures, vertical autorotations, rough trail operations, and jump takeoffs.

Meanwhile, when we weren't flying ground school, physical training, parades, and inspections continued to be part of the schooling. It was a busy schedule, so busy that one week I forgot to get a haircut for the Friday inspection. What do I do? I conned my protesting wife to trim the hair on the back of my head and neck and around the ears. I figured my hat would cover any other problems. I couldn't see what she did.

The commanding officer of Ellyson was the inspector. When he walked behind our row of the formation, I was holding my breath, standing ramrod straight and staring straight ahead. Then he stopped and studied the back of my head.

"Who cut your hair, mister?"

Startled, I blurted out, "My wife, Sir," and gulped.

Then he made a few derogatory comments I couldn't understand. When he finished the inspection, he stood in front of all the students in formation and announced, "I don't want anyone's wives giving haircuts."

And mine didn't again.

I graduated from basic helicopter training and advanced to the more sophisticated HUP-2. It was the second generation of operational rescue helicopters assigned aboard Navy ships. The tandem rotor HUP was truly monster in size and especially in how it was operated and controlled compared to the small and simple HTL. It became a Navy legend. The original fleet rescue helicopter was the Sikorsky HO3S—the helicopter of Michener's of *The Bridges of Toko-Ri* fame.

The first HUP went to Helicopter Utility Squadron 2 (HU-2) and entered fleet service on the East Coast in 1951. It was built by the famed helicopter pioneer, Frank Piasecki. It was 32 feet long with three hollow rotor blades attached to articulated rotor heads at each end of the airframe and was powered by a Continental R975-46A radial air-cooled gasoline engine. It initially produced 650 horsepower at max power, but was later derated to 450 horsepower to lower the engine failure rate. It was the most underpowered helo ever designed for the Navy.

We were told the engine was a modified tank engine from World War II, and Piasecki was told to use it to build 338 of these helicopters for the Navy, U.S. Army, Royal Canadian Air Force, and French Navy. There had been modifications to the engine to provide for better oil flow, but the engine still had an enormous failure rate.

Reciprocating engines were not designed to operate at near-maximum rpm and have a long life. One difference between today's turbine-powered helicopters and those reciprocating engines of years past is that the turbines can operate for long periods of time at maximum power and do so efficiently.

Although I didn't realize it at the time, the HUP and I were to set several records during an exciting 2-year period together. It was a different and much more complex machine than the HTL.

The HUP weighed 5440 pounds at maximum gross weight. It could carry enough fuel for about 4 hours of flying time plus a pilot. With less fuel, it could fly with maybe two or three passengers depending on weather conditions. It required more movement of the controls to get flight control results than the lightweight HTL. Its initial top speed was 120 knots with help from an autopilot, but the autopilot caused some major problems and tragedies, so the forward main rotor head was tilted back about 3 degrees to reduce maximum speed to about 80 knots.

My first flight was just to get a feel of the machine. The very first lesson I learned was to never let go of the controls. With the HTL you can tighten down the collective or at least hold it with your left knee on a straight and level flight and be able to light up a smoke.

The HUP required both hands and both feet at all times, and the pilot had to make constant corrections just to keep the machine headed in the right direction at the right altitude.

"There is a big difference between the HUP and all other helicopters," warned Lieutenant Nelson, our transition instructor.

"One of the big differences is the light weight rotor system," he explained. "If you have an engine failure in the HUP, the hollow wood fabric covered rotor system slows down quickly since there is not enough weight to give it inertia to keep spinning," he added. "So get that collective down fast and keep those blades windmilling. Then the tricky part comes when you start to overspeed the rotor system. You can't hold the collective down for the entire autorotation from 500 feet or higher because the rpms will build up so fast it'll sling off all the blades due to the centrifugal force. So keep an eye on the rotor rpm and lift up on the collective so the rotor blades take a bite of air and slow down to keep within proper rotor rpm limits while you are descending.

"Depending on your airspeed, you'll have to make constant collective adjustments. During our practice autorotations you'll also have to keep the engine power up so we can marry the needles about 20 feet in the air to execute a waveoff. We don't want to make a touchdown during an autorotation with this helo. We've been banging up too many tailwheels, which puts the helicopter out of service for weeks while repairs are made.

"Good luck on your landing. If you ever have a real engine failure, all you do is pull up on the collective just before you hit the ground. The blades will take one bite into the air and act like a skyhook—then hopefully you'll settle to the ground. That is something we just don't practice.

"Now let's go fly this beast."

As we lifted off for my first hover, I was working both rudder controls and the cyclic continuously and there was a terrific buffeting.

"This monster wants to wander all over, and what's causing the buffeting?" I asked Lieutenant Nelson over the ICS (intercom system).

"Oh, that's caused by the forward rotor blades turning clockwise and the rear rotors turning counterclockwise, which forces all the air created by the rotors to pound against the right side of the helicopter, creating lots of turbulence. You'll master it," he said casually. And I eventually did, but it took awhile.

"Here we are at an altitude of 500 feet heading straight for the square," Nelson told me after a few hours in the HUP. "Now split rpm needles, and let's see you autorotate to the square on the practice field."

I pushed down on the collective, the helo dropped so fast my stomach was left behind. I couldn't even look out the canopy, I was so worried about adjusting the collective to keep the right rotor system rpm speed. I didn't want to overspeed it or have it cone either.

Then I noticed on the rate of climb or dive indicator that we were plummeting at the rate of 2000 feet a minute. It didn't take an Einstein to figure out we'd be on the ground within 15 seconds. It seemed to me that was a sink rate faster than a rock.

"Come on, McKinnon, watch where you're heading," Nelson reminded me. "Try to forget the instruments and look in the direction we're headed.

"Here, I've got it," he commanded as we were approaching the ground. His hands were on the controls right along with mine, except now he took charge, married the needles, and saved us from hitting the ground.

I couldn't believe how fast the HUP sank. I was way behind it. As time passed, I got the hang of it. After 8 hours he let me solo. It was an uneventful ½-hour flight. I never left Ellyson Field and was exhausted at the end of the hop, more from worry about what might happen than what was happening.

Mastering the HUP became a challenge, and after 28 hours in this helo I felt comfortable. Then my orders came.

There were 12 helicopter squadrons in the Navy, 10 antisubmarine warfare that dipped transponders in the water to search for submarines, and two utility squadrons—HU-1 and HU-2. The helicopter pilots I admired during my midshipman cruise were a detachment from HU-2 located at Lakehurst, New Jersey—HU-1's East Coast counterpart.

I received my designation on December 3, 1957, as Navy helicopter pilot 3823 and orders to the Navy's first and most famous helicopter squadron—Helicopter Utility Squadron One—HU-1 at NAAF Ream Field just south of my hometown, San Diego. Another stroke of good fortune.

CHAPTER FIVE

Helicopter Utility Squadron One

WARM AND SUNNY SAN DIEGO is a great place to be in the middle of the winter. But even better, it was great to be there as a full-fledged naval aviator assigned to an operational squadron.

Before reporting to HU-1, checking in with old friends was in order, so I dropped by NAS Miramar in full uniform, including the shiny gold wings, to visit my old friends at Operations—friends I hadn't seen in 15 months. Chief Simmons was there. We talked about flight training, and he was surprised, yet delighted to learn that I had become a helicopter pilot.

"Let's jump in our HUP and go for a ride," he offered.

"Great!"

That was the best thing he could say to me. I didn't have a flight suit with me, so I just took my khaki uniform jacket off, borrowed a helmet, and we were off. He now treated me as an equal.

"Here, get in the left seat," he offered.

It didn't take me more than a second to accept. I got the helo started and blades engaged without any problems as I showed off my newly learned skills. We had a good time chasing around, up and down the dirt fields north of Miramar and practicing maneuvers at Hour Glass Field. After 45 minutes of low-level flight, we returned to Miramar. He let me make the landing in front of Operations. It was an acceptable landing, but not so smooth to avoid kidding from the old gang I had once haunted with questions a little over a year ago. I now really felt like I belonged to a special group.

NAAS Ream Field was located in the southwesternmost tip of the United States, bordered by Mexico on the south and the Pacific Ocean on the west. It was a 246-acre auxiliary airfield for NAS North Island. It had been a base since 1917. All the structures were old, temporary, single-story buildings. They were made of material one grade stronger than cardboard—all leftovers from World War II.

The airfield was home to four helicopter squadrons in addition to HU-1. Those included HS-2, HS-4, HS-6, and HS-8, all with antisubmarine sonar-dipping HSS helos. Most of the pilots had originally been members of HU-1. Now they were stuck with what I thought was a routine and boring type of helicopter flying.

Also stationed there was an outfit flying World War II F6F fighters. The F6Fs had been converted to remotely flown drones. They were launched from Ream and flown to sea, where they were used by Miramar fighters and ships at sea for target practice. It hurt every time I saw one of those great World War II fighters head out to sea to be shot down and destroyed. What a waste of a great old warbird.

Launching the drones from Ream Field was a safety feature. They'd climb up from the runway and be over water immediately so that the civilian population would be protected from a crash. There also was an approach over unpopulated areas in case the drone wasn't shot down in training and the escort plane would fly it back to Ream for a remote-control landing.

HU-1, the squadron known as the "Flying Angels," was located on the east end of the field in a series of old parallel buildings. Our squadron helicopters were parked on the ramp in front of those buildings.

"Lieutenant jg McKinnon reporting for duty," I told the squadron duty officer and gave him a snappy salute.

"Welcome aboard, Lieutenant. We've been expecting you."

"You have?" I questioned.

"Yep, we keep track of all the new arrivals. This is a big squadron with nearly 100 pilots—we have at least a half dozen coming and going every month. You'll never get to know every pilot in the squadron because we'll always have about a dozen units with from one to four pilots deployed on 6-month cruises at any one time. Your orders say Clinton Dan McKinnon. You go by Clint or Dan?"

"Dan."

"Dan, you're joining the Navy's first and oldest helicopter squadron. It's got a lot of history since it was first commissioned on April 1, 1948. Here's your checkin sheet. It'll help you to get to know our squadron as you get it signed off at each department."

I started making the rounds—Administration, Personnel, Medical, Maintenance, Operations, and XO (executive officer)—and finally met the CO (commanding officer). The CO's yeoman had his desk shoehorned into tight confines in the area of the executive officer and commanding officer. I told the third class petty officer I was here to see the XO as part of my checkin. He knocked on the open door and announced me.

"Tell him to come in," boomed the XO's voice from a small office around the corner.

"Welcome to HU-1, Mr. McKinnon," said Commander C. N. McKenna. A tall man with a large lanky frame ambled over to the door and jutted out his huge hand for a shake. My hand was lost in his palm. He gestured me to a seat in front of his desk covered with stacks of papers.

"We've been expecting you. You ready to start flying?"

He didn't need to ask that question, I thought to myself. Of course, I am.

"Yes, Sir," I told him without hesitation, looking him square in the eye. Then he gave a little lecture about my squadron.

"We'll try to get you on the flight schedule as soon as you get finished checking in and get through survival training.

"We have our own flight training program. You'll have to finish that before we can send you on a cruise. You learned a lot at Pensacola, but flying off ships is demanding and requires skillful flying. We want you to get so you can fly a helicopter with exact precision. Errors result in accidents. We don't want you to have any problems aboard ship. Helicopters already have the highest accident rate in the Navy, although our fatality rates per accident are less than the jets.

"HU-1 has a variety of missions, and the most exciting and dramatic is rescues, but we also transfer personnel, cargo, and mail between ships at sea; fly planeguard; spot for Naval gun fire; search for mines; fly photo hops; evacuate wounded; orbit for radar calibration; transfer VIPs between ships and historical meetings; track torpedo launchings; plus demonstration flights to show what our helicopter can do. We also help with expeditions in the North and South Poles with Operation Deep Freeze projects, including ice reconnaissance, to name a few of our missions.

"We operate the HTL, HUL, HUP, HRS, HUS helicopters plus the SNB to keep instrument-qualified. We have 102 officers, all pilots, about 530 enlisted men, and 40 helicopters. Since the squadron was first founded nearly 10 years ago, we have made a total of 648 rescues. You're going to be assigned to fly the HUP.

"Once we get you trained, you'll probably make a long cruise, 6 months or so, flying planeguard aboard a carrier. The long cruise deployment is the payback for all the training and investment the Navy has in you as a pilot. Any questions," he concluded.

"No, Sir."

"The way we're set up around here, all pilots have collateral duties, besides flying. Tell me a little bit about your background and what you majored in in college."

"Well, Sir, I majored in history, political science, and a little economics. I grew up working at a newspaper after school and my dad owns the Sentinel newspapers. I also used to take a lot of pictures."

"Our PIO [public information officer] has just left on a cruise. You'll be our new PIO," he said without further discussion.

"Yes, Sir."

"Let's see if the skipper is in. He'll want to welcome you."

He went to the adjoining office and came back in a couple of minutes.

"Go on in, McKinnon."

"Sir, I'm Lieutenant jg McKinnon checking in, Sir," I told a medium-built rather stiff man standing behind a fair-sized desk.

"Relax, sit down," Commander Al Snider said rather coldly. I could tell immediately he wasn't the kind of man you'd relax around. He had piercing black eyes and a dark face with a perennial 5 o'clock shadow, even immediately after shaving. He lacked humor or warmth.

He had just taken over command a couple of weeks before. He had assumed responsibility for one of the unique squadrons in the Navy. His pilots were scattered from the North to the South Poles, from San Diego to the Far East aboard every carrier, cruiser, and even some LSTs, plus a detachment that serviced the helicopters in Oppama, Japan (near Yokosuka).

It probably took 9 months before he ever met all the men who reported to him, and he had only a 1-year command. This was far different from the fighter squadron of 14 men and 12 airplanes working and playing together as a team. In my 2 years assigned to HU-1, I never met all the pilots, either, there were so many coming and going on different assignments.

"The XO tells me he has briefed you about the squadron. Safety should be paramount in all you do. We don't want to lose any pilots or planes for that matter," he emphasized.

"This type of squadron needs mature pilots. You'll be operating with just a few pilots or by yourself for long periods of time, and we need the type of officer we can trust. You may end up even being your own boss. We're known for rescues, but the chances are that only 25 percent of the men in the squadron will make a single rescue during their Navy careers, so good luck. I hope to fly with you one of these days soon."

That ended the checkin process. It was strange. I only had about six other face-to-face meetings with my commanding officer until I left on my cruise 6 months later and never did fly with him.

After passing a routine flight physical, I was cleared to go flying.

I stumbled into another valuable lesson during the process. We had periodic physicals, shots, and other reasons to go to sick bay to get a flight surgeon's attention. There were always lines, long lines, even for officers. What a waste of time, I thought.

Then one of the doctors asked me how he could get in some flight hours for his flight pay.

"No problem," I quickly responded.

The next day I was scheduled for a training flight. I got a hold of the Doc and had him sitting in the backseat. I gave him a running commentary throughout our hop. I saw to it thereafter that he always got scheduled for his necessary flight time. And in turn, I never waited in line again for anything at sick bay.

The training program at the squadron was the same, yet so different from that in Pensacola. In some respects it was like learning to fly all over again. The squadron instructors demanded that everything be so precise. When you made an approach to a point, you were expected to come in right to the point. In Pensacola we could come in 10 or 20 feet off as you gained control of the helo. At HU-1 you hovered or tracked along the lines of the squares in the practice area. You learned to do so exactly along the line and not vary your altitude more than a single foot.

Shipboard operations required meticulous control. The squadron instructors were hand-picked. They flew operational missions, but were also required to give demanding training to all the newcomers. All had successfully completed a long cruise. They knew what the requirements were for a shipboard-qualified pilot and had the ability to communicate, demonstrate, and impress on the newcomer what was important.

Once again my piloting skills reached a new level, and I looked back when I imagined that I was so great. I understood now that I had to reach a higher level of expertise. The cycle of learning never seemed to end.

When not flying, my collateral duties as PIO (public information officer) absorbed all my time. It was a fun, exciting job that always placed me in the center of activity.

If our squadron flew any unusual flights, the press wanted to know. All civilians contacting the Navy about helicopters were referred to our squadron. HU-1 was the center of activity and diversity. It made good press. The HS squadrons at Ream practiced hovering and lowering their sonar domes attached to long cables under the water to find, plot, and track submarines. Two sonar crewmen would sit in the back, listening for pinging echoes above the rotor and engine noise. The pilots would trade off hovering until they returned to the carrier. It was almost as boring a job as flying a transport plane.

Our squadron was able to romance the press about the new contraption called the *helicopter* and all its versatilities and capabilities. I was assigned an assistant, Chief A. J. Kress. He was about 20 years my senior in both the Navy and age. He knew just how to handle a new, young naval officer. He made me think I was important and in command, when in reality he made most of the decisions and did the bulk of the work. We actually made a great team. I usually had a lot of public relations ideas and then would go off flying and Chief Kress implemented them.

With detachments scattered all over the world, our squadron had a hard time communicating. As PIO, one responsibility was putting out a monthly newspaper called the *Chopper*. It was circulated to each member of HU-1, both at the squadron and to deployed units all over the world. It also went to cruise "widows" so they'd have some idea what was going on in the squadron. We had each squadron department write about events that affected them and included general gossip, plus a required report from each deployed unit.

With 10 or 12 units deployed aboard ships on extended cruises all over the Pacific Ocean, it wasn't just informative social news, but a report of operational problems encountered by the deployed units that served as an education outlet for all. There was little Navy-wide publication in those days that shared operational problems. *Approach Magazine*, the Navy air safety magazine, was only 4 years old.

Those of us preparing to deploy scoured those reports to glean all the information we could. The real genius behind the newsletter was Chief Kress, who had been with the squadron for a number of years. He really put it together with a little help from me. He had a folksy knack of writing stories and took great pride in the newsletter. He really made me look good by taking a special interest in publishing the *Chopper*. It was difficult for me to keep a consistent schedule in the office because I was off flying or taking short cruises as part of the buildup process for my long deployment.

On April 1, 1958, we had a special celebration and cake-cutting party for the 10th anniversary of HU-1 as the Navy's first operational helicopter squadron. In the first 10 years of operation the squadron had made 657 rescues.

We also sent out news releases about every rescue we made and had news stories in the San Diego Navy newspaper, as well.

Flying and having a collateral duty was demanding at times. The Navy had a system of assigning pilots to desk and office jobs as well as flying. This helps reduce squadron-manning levels for the cramped areas on carriers. The Air Force guys with their more spacious accommodations only had to worry about flying with no responsibilities for the operations within the squadron. Those were handled by what were termed *nonrated officers*. Inside the Navy we complained about it; outside, especially when talking to Air Force types, we bragged how it made us better officers with greater leadership experience.

We also put together a weekly series of stories featuring the squadron in a local Navy newspaper. And, of course, we gave tours for local groups and schools, plus talks to Rotary (Rotarian) clubs and other organizations.

One major project I came up with was to replace the squadron insignia. It was an old-fashioned fat bird replica made into a helicopter with a couple of miniature wings on the side. There was a lot of nostalgia to keep the 8- or 9-year-old emblem, but it didn't match up to a more sophisticated image that was currently being displayed with the expanded role of the squadron.

The purpose of the insignia on plaques, decals, and patches was to foster morale, esprit de corps, and a sense of unity. It was used to symbolize the mission of our squadron.

The XO agreed to a contest within the squadron to pick another design. A free trip to Las Vegas, donated by a major hotel, in exchange for some free publicity would be the winner's prize.

The winning design exemplified our mission. It was a circle emblem of a helping hand reaching down from a heavenly cloud through a halo with attached wings, signifying an angel. It had a hand grasping up from the choppy seas searching for help. The backdrop was the Pacific Ocean with the U.S. shoreline on the right and countries of the Far East on the left, which signified our operating area. At the bottom was a ribbon with the colors of the presidential unit citation, which had been awarded to the squadron, and the words *Hutron One*—abbreviation for Helicopter Utility Squadron One. That was later changed to *Helicopter Combat Support Squadron One.*

It was designed by one of our own squadronmates, a first-class mechanic out in the line shack named Bob Matthews. He did a great job. The CO and the Pentagon eventually approved the design, and it was still with the squadron 36 years later when it was decommissioned.

After 4 weeks with the squadron, another part of my training involved orders to survival school. In addition to teaching basic survival techniques, it was an ordeal of physical torture and mental anguish to teach us to withstand the pressures of torture and brainwashing behind enemy lines, should any of us become a POW.

CHAPTER SIX

Survival School

THE FIRST THING we faced on arrival at Survival School early Monday morning was a strip search—not to simulate being a POW—but to find and take away contraband each student intended to use to supplement his eating habits during the next week.

We'd all heard stories of how tough the week would be and everyone decided to smuggle supplemental supplies to augment what we believed would be a starvation diet at SERE (survival, evasion, resistance, and escape) training.

The instructors had a field day—candy bars, raisins, nuts, instant coffee, tea, hot chocolate, beef jerky, and bouillon, and dehydrated soups. You name it, they found it in the pockets of our flight suits and jackets.

The confiscation was routine for Survival School instructors, but for the students it was a big deal.

All those snoopy instructors allowed us to keep was what we might ordinarily carry in our flight suit on a typical mission. The goal was to simulate conditions of being shot down or crash landing in enemy territory and how to respond. It was intended to be realistic as possible.

My two packs of five flavored candy Life Savers were considered okay. They also let me keep a handful of waxed matchbooks, a pocket compass, my hunting knife in a sheath sewn to my flight suit, and a partial roll of toilet paper. Even though no one normally carried toilet paper in their flight suit, they let those of us creative enough to think of it to keep it. That paper was to become my most precious possession.

The guys who flew the big patrol planes, like the P2V *Neptune*, really let out a holler. They claimed since they flew long 6- and 8-hour flights, they usually carried plenty of supplies in their flight coveralls. They yelped and argued that they weren't being treated fairly. The instructors turned a deaf ear; they showed no sympathy and made sure those crews realized life wasn't fair.

The instructors must have been tired of hearing the same old excuses, but once in a while they were impressed with the ingenuity of an occasional student.

The school started at North Island Naval Air Station and concluded at an isolated POW compound near Warner Hot Springs in the Cuyamaca Mountains, 75 miles east of San Diego.

We had a class of about 50 students—roughly 20 officers and 30 enlisted men. Squadrons from all over the West Coast with personnel who might find themselves in combat or behind enemy lines were represented. There were only four of us from HU-1—another new arrival, Lieutenant jg John G. Loomis, and two enlisted flight crewmen.

Our chief instructor and man in charge of the school was a tall, large, burly-chested, tough, thick-skinned lieutenant commander with bushy gray hair. He had a charisma and presence that dominated the room.

His name was Bob Chilton. He was a legend already. He eventually would train more than 35,000 students before being forced to retire in a Navy cutback.

Chilton flew as a World War II and Korean fighter pilot and had four air-to-air kills. But his eyesight had faltered and he was grounded. He was looking for something fulfilling and found his niche and a new purpose to life by helping train and motivate flight crews with physical and mental survival techniques. He once said it was a lot more rewarding to teach men how to live rather than to teach men how to kill.

Chilton was the kind of guy who demanded respect as he strode around in his open-collar khaki uniform, cigar chomped between his teeth, yet talking with a forceful and booming gravelly voice. Although this school was only a 6-day event, the 10 or so hours I spent that week listening to Chilton's lectures on survival, mental toughness, and importance of self-discipline left an indelible impression on me.

It would last a lifetime for me and thousands more.

I couldn't get enough of Chilton's intensive lectures.

Being in the rescue business and especially with HU-1, I knew if war came again, there'd be an excellent chance I'd find myself behind enemy lines trying to pick up some downed American pilot—just like Mike Forney played by Mickey Rooney in the *Bridges at Toko-Ri*. Only I'd planned to be successful. That thought alone, was plenty of motivation to listen and learn.

Chilton had the ability to stir your emotions. He caused me to determine if I ever found myself in a survival situation, I was going to have the mental toughness and will to survive and escape. He made me believe that if I ever went down, I would be prepared to do whatever was necessary to escape capture.

I'd done a lot of camping and backpacking in the High Sierras, including climbing Mt. Whitney while in high school. So I had a natural interest in the outdoors and a knack for improvising to take care of myself with whatever I had in my pack. But Chilton lit the motivation fire of interest to elevate a person's thinking from living in the outdoors—to surviving in the outdoors during adverse conditions.

He would remind us that there was a big difference between survival and a survival situation.

Chilton took care of housekeeping chores first.

"This week the chain of command is in effect," he declared. "If you want anything other than to ask a question in the classroom, ask your senior officer present. This is the way it works in a prison situation and that is the way it works here.

"There's no sleeping in the classroom. If you get sleepy, stand up.

"Be prepared to be wet and cold all week. There is no such thing as hot water this week. You're going to eat less than you are accustomed to.

"You'll be divided into groups. Each group eats its own food. There is no community pot. No one is allowed to use paper to start any camp fires.

"Anyone caught using contraband food or violating any of the rules of the school will be returned to his squadron for disciplinary action.

"We're going to teach you how to deal with pain, heat, cold, thirst, hunger, fatigue, fear, boredom, and loneliness. When you're under psychological stress we want you to keep cool when things get hot.

"One airman out of three who bails out leaves his plane uninjured but is hurt or killed on landing. Those injuries are caused by smashing into trees, drowning under the chute, poor posture in ejecting, wrong body contact on landing. So know your equipment and how it works. Whether you are injured or not will make a difference if you become a POW."

"In Korea," Chilton said, "official figures showed 38 percent of the American POWs who reached an enemy camp died in captivity. There were 7140 prisoners taken in Korea. The missing in action figure was over 11,000. The gap of almost 4000 is unaccounted for, so we know more than 2701 POWs died there.

"During World War II, in Japan, the POW death rate was 38 percent. The death rate of American POWs during World War II in Europe was only 2 percent.

"In Korea malnutrition was the basic cause of death in the prison camps. That lead to many diseases, the worst being pneumonia. With no antibiotics it was 100 percent fatal.

"So eat any scrap of food you get possession of whether you like it or not. Don't wait till later. Keep your strength up. No telling when the enemy will provide food next. Your body needs it. We're going to teach you that lesson this week.

"Another direct result of malnutrition is pellagra—a disease of vitamin deficiency which causes diarrhea and dementia. Dementia is apathy.

"During 1950 and the spring of 1951 such vitamin starvation in POWs induced a 'give-up-itis' or 'face-the-wall' syndrome.

"The only technique that worked to save lives then was to harass, gloat, prod, or hit to get a man to his feet so he'd get mad enough to try and beat you up. If you could get the suffering POW so mad he would get out of bed and declare, 'if I could get my hands on you I'd kill you,' the man invariably got well.

"Evoking such anger saved many lives in Korea. If you're ever in that situation, take whatever actions are necessary to save any of your fellow

POWs. Say any words or thoughts necessary to taunt a POW in that condition to get him angry and save his life.

"The Korean War gave us the first wholesale use of a POW for propaganda in the form of confessions. There were germ warfare accusations, war crime confessions, and propaganda.

"Each man must realize the enemy wants to use him—this realization will aid in his resistance," Chilton instructed.

"No matter how severe the treatment you receive, you should resist to your maximum limit to retain your personal pride.

"In the next couple of days we'll teach the Code of Conduct and Geneva conventions to you.

"During the Korean War a new type of interrogation and mental torture took place and changed the way America now approaches POW training and the need to train POWs to mentally resist.

"Now many of you think you have problems, personal problems. Let me draw you a picture of every man in this room," Chilton said smugly hoping someone would challenge his artistic ability.

He walked over to the blackboard, took a piece of white chalk, and drew two long horizontal parallel lines about 3 inches apart nearly the length of the blackboard. He drew a circle at each end of the two lines. It looked like a long cigarette or tube.

"You recognize yourselves?" he queried.

"Imagine this is 36 feet long, and this is the end the food goes in and this is the end it comes out. If you take care of one end, the other will take care of itself."

It became obvious that he was talking about our digestive tracts.

"But if no food enters this end for 72 hours," he said pointing to the mouth end, "what's going on in the rest of your life, whatever problems you think you have disappear, until the mouth is satisfied.

"In a survival experience the mouth will become an overriding problem to fill and will become your number one problem in life. That's something you now take for granted in this rich country of ours."

Following our first classroom session Monday morning, it was off to the swimming pool to prove our abilities in the water. There were endurance laps in flight suits with waterlogged boots on, underwater distance swimming tests, jumps off high platforms to simulate abandoning a sinking aircraft carrier, and even swimming tests simulating that we were escaping through water with ignited fuel burning on the surface. It was demanding, physical exercise, and some in the class needed remedial sessions to pass the test requirements.

The instructors stuck to their word—no lunch after all the strenuous workout. The big breakfast I had stuffed down was keeping my energy level up, but the coffee and donut guys were starting to suffer. It wasn't just going without food—it was going without food and knowing that there was no hope of getting any until we scavenged some that evening on the beach.

The psychological effect was powerful.

The idea of weakening our resistance and affecting our judgment through hunger and exhaustion started to enter our minds.

At the end of the first day of lectures and swimming tests, we were issued half a parachute to use as a sleeping bag plus one full parachute per group of five men for a tent. It took some practice to learn the special technique of folding the partial parachute to make it a warm sleeping bag. We also had to learn how to put together the parateepee with three poles of bamboo to make a tent. We were instructed to keep our fingers from touching the chute. When the night dew came, any place touched on the tent parateepee would leak water on us all night while we tried to sleep.

But the real challenge was eating—or not eating.

We were bussed to the sandy beach on the south side of North Island, bid farewell by most of our instructors, and admonished not to leave the area. They'd be back for us in the morning.

"What about food?"

"You're on your own! See you in the morning."

The winter sun was setting. All our overnight preparations, search for seafood, and camp setup had to be accomplished in the dark.

Enthusiasm ran high to show those SOB instructors we could take care of ourselves in spite of their doubts and scorn.

My small group of five divided into work parties. Some gathering driftwood and building a fire and others going out on the North Island jetty that was a breakwater between the Pacific Ocean and San Diego Bay. We clumsily climbed around in between the giant slippery rocks of the jetty and fished, grabbing with our hands. We carried old used metal 5-gallon containers to hold our catch, which consisted of large crabs that we were able to capture between the rocks.

We never caught a fish.

Occasionally someone's foot would slip on the algae-laden rocks and he would slide into the water, soaking his boots, socks, and flight suit. But we returned to camp with several buckets of crabs.

The metal containers were perfect for boiling the crabs in saltwater.

By the time we finished boiling them, everyone was dirty and smelled of smoke. The most tender part of the crab was the giant claw with grainy white meat that seemed to break into pieces the size of grains of rice. We picked them apart with our hands. No silverware or utensils this week— just our hunting knives. Any fussiness we may have had about dirty hands or dirt in our food faded. Our growling stomachs made sure of that.

At least half the guys swore they weren't going to eat any of that stuff—and didn't.

It was a miserable night. The sea breeze blew the sand around. It rubbed and chafed us in our flight suits and got in our eyes and hair. The saltwater from falling off the jetty rubbed some of us raw and kept us cold. We really didn't get our parachute sleeping bags folded right. The sandflies nibbled on us and we shivered most of the night. Some of the tents collapsed from the wind, and by sunrise most of us were grumbling and anxious to get going. No baths, showers, or shaving. We handled toilet requirements like dogs along the beach.

At 0700 hours Tuesday the buses arrived to shuttle us back to the classroom for our morning lecture.

The coffee mess was off limits. Jitters set in among the coffee addicts. Morale wasn't exactly what it was the night before. We began to realize this wasn't just a fun campout.

Chilton's lecture started off with a hard-hitting approach. He attacked our current lifestyle to stimulate our thinking about mental toughness.

"Most Americans have never had a hand laid on them in violence," he snarled. "Most of you have been raised by parents who provided loving care—then suddenly, if you're unlucky and you become a POW, somebody not so loving has his hands on you, and is in command of your life.

"You've grown up soft," the Kentucky outdoorsman and former boxer emphasized with intensity.

"You've picked the cherries and let the bears have the trees.

"It started in school. There's a total lack of discipline today. It's self-discipline that's going to get you through life—and a prison environment. You're going to have to pay the price to find out what you're worth. To find out just how tough you are. You'll never know what freedom is until it's taken away from you. And we're going to take it away from you this week."

I sat there riveted to my seat by his comments and wondered what was ahead for me this week.

"You're living a soft life. Wanting to buy everything on credit—not only in what you possess, but wanting instant gratification. As a POW you're going to be in for a rude awakening.

"You need to work on your self-respect," he lectured.

"How do you get self-respect?" he asked.

"You achieve something, anything, through your own efforts," Chilton answered his own question before anyone in the class could respond.

Chilton had style, drama, and a magnetism to capture everyone's attention.

I hung on each word, wrote voluminous notes, and tried to inhale the feel and emotion of his comments.

I shared his beliefs.

We Americans were fortunate to have everything so soft in this country, especially compared to the rest of the world. We needed to mentally prepare ourselves for the tough times that could lie ahead for our country.

Chilton just wasn't talking about life as a POW, he was talking about living life every day. He tried to get us to focus our goals and aims as a person.

"Concerning survival, think of it as a way of life. We have forgotten in this modern-day age of plastics and can openers, some of the basic things our own forefathers, the pioneers, used in everyday life.

"Survival in the purest sense stems from one's ability to stay alive, sane, and productive in an environment which 'hands out' nothing to the individual. Many people in our country have faced survival situations in the past.

"It seems certain there will be more and perhaps even greater challenges to our way of life in the future," he continued. "The measure of success is found in your attitude as an individual. When you joined the military, you became legal game. You've got a big target on your front and back. Nobody told you that.

"As a POW you'll be stripped bare of all your outside possessions. All you'll have left is your character.

"What are you really like on the inside? You'll be jerked up off your feet, slapped in the chops. Most people have never lost control of their situation before. Most of us have never been completely helpless while someone else dictated the terms.

"The shock of that will have an effect on you. You have to take what comes. Your self-reliance will surface—you'll find out what inner strength you have.

"You must think positively—if you think you're going to die—you most likely will."

I couldn't tell whether he was talking about this week or if we were ever a real POW, but either way it sounded like good teaching.

"Some people should never be in the military. Some people can't train themselves to take the fire. Some people just can't take it. The truth is, in combat such a person is a danger to you.

"Only 10 percent are going to win the war. The rest are on the outer edge of the fight or flying too high to be effective in getting the fire power on target.

"Someone could be flying wing on you and be such a weak link he could get you killed because of not being able to function in combat.

"That boy shouldn't be there—it's not right to ask him to be there. In America, we think everybody should be a soldier. Not everybody can be a soldier.

"We'll find a weakness in you. We'll get it close enough to the surface. We'll get you to the edge so you better understand yourself. We don't want to break you, but we want you to understand yourself so that if you become a POW, you'll be able to cope."

Chilton went into detail on how to behave if captured, particularly by Orientals since Far East countries like North Korea and China represented the biggest threat to peace and stability for the United States and were the areas we'd operate in as Pacific Fleet pilots.

The Korean War experience had truly opened the eyes of Americans to the brutality generously meted out by Orientals to American POWs.

"Seldom before Korea did the bulk of men taken prisoner have to endure the severest conditions of captivity—like they did in Korea. There it was a novel blend of leniency and pressure with rare outright cruelty.

The three psychological effects used most frequently were repetition, harassment, and humiliation. What, then, is brainwashing?

"It is a word that has become embedded in our language that refers to communist prison overseers who coerce, instruct, persuade, trick, train, delude, debilitate, frustrate, bribe, threaten, promise, flatter, degrade, starve, torture, isolate individuals or play off members of a group against one another to suit their purposes," Chilton rattled off in quick succession. "In essence, a prolonged psychological process designed to erase an individual's past beliefs and concepts and substitute new ones.

"The most frequent reason given by prisoners in Korea for things that went wrong was their total lack of preparation for what they encountered," Chilton explained.

"They claimed they had no idea of what to expect. They were given no idea of how to handle that kind of situation. They had no instructions on how to act as prisoners. No one discussed collaboration with the enemies or going into indoctrination sessions.

"Illness, death, disorganization, ineffective escape attempts, and effective harassment by the prison guards haunted the POWs.

"If Army personnel captured during the Korean War had been given so much as a minute's instruction on what to do in the event of capture, or even the measures for evading capture when cut off from their forces, it would have made a difference. That's why this school exists, to share the lessons of Korea," Chilton concluded before a break.

True POW stories were shared by Chilton. He asked us to think about the sequence of events those men encountered, to realize words that could never completely capture what really went on, so as not to set ourselves up as judges on the subject of those stories. Instead, he wanted us to evaluate a sequence of behavior in terms of the actual consequences and in terms of alternative behaviors which might have been available to the person who suffered.

He wanted us to think—given the same set of facts—how we might have responded.

The case study approach forced us to systematically analyze ourselves, our own thinking, our value systems.

What procedures would we use as individuals?

It all began at the point of capture. Then he walked us through the phases, such as exposure to local populace, being protected from them, transportation with our belongings, the results of interrogation along the way while being transported to some central detention facility.

What conclusions could we draw about a particular enemy organization, communications, and discipline?

What were their objectives with the POWs?

He concluded by noting several principles:

1. Strong family ties and/or a strong faith are two essential ingredients to weather the storms and trials of being a POW.

2. Strict self-discipline is a key to survival.
3. Care for others brings out strength in you. It's very difficult for a man to give up while he is busy helping others.
4. Most successful survivors of POW camps either had or developed a wonderful sense of humor.

The afternoon pool session Tuesday was spent with more swimming tests. We ignited daytime orange smoke flares and nighttime red flares, and learned the makeup of emergency food rations and how to work salt-water stills. We learned about survival kits and how to make a PSK (personal survival kit).

The Navy food rations were stored in waterproof containers about twice the size of a can of spam. We even got to eat what we wanted of the K rations, including dry meat and cereal bars during these introductory lectures. The taste of those cellophane-wrapped cereal bars was unforgettable—they looked like compressed sawdust, but sawdust would have tasted better.

We weren't hungry enough to appreciate the rations yet.

We also learned how to be hoisted into a rescue helicopter. We learned that if you entered the rescue sling correctly, it was virtually impossible to fall out. If you entered backward, you were assured of being hoisted half way to the helo, falling out and doing a face-first belly flop back into the ocean.

Tomorrow was going to be our test. Helicopters were going to lift us as we paddled in the ocean.

One great feature of all the poolwork was the freshwater bath. At least we got rid of the sand and saltwater from the night before.

It was our second night on the beach. We had the routine down much better. One of the guys even captured two giant lobsters. Many of us thought it a great treat. There wasn't enough for everyone, but those who liked lobster enjoyed each little morsel.

There was another change. At least three-quarters of the group ate crab. There were still some holdouts who refused to eat what others considered a delicacy, but as the few candy bars and other food items allowed were consumed, the starved troops showed interest in food they didn't like.

We had a better understanding of how to prepare our camp. We got the parachute sleeping bags folded properly and erected more stable parateepees. We started learning a technique on how to search more effectively and snag the crabs with our bare hands. We adapted to sleeping on the hard sand with no mattresses, or was it just exhaustion?

John Loomis, the other HU-1 pilot, and I began to work as a team. There was something about sharing a rough experience with someone else.

Being really tired after 2 days of vigorous exercise was a real aid to sleeping better. We were up at the crack of dawn again Wednesday.

In preparation for Thursday's cross-country mountain navigation hike to the Cuyamaca Mountains and Friday's escape and evasion course, the lecture Wednesday covered Code of Conduct, Geneva Convention, and basic evasion techniques.

"The first half hour in enemy territory is the most important," Chilton lectured.

"And it's your greatest opportunity to be rescued. The enemy will be after you. You'll probably have landed or parachuted in a remote area from the enemy troops, but your parachute will have undoubtedly been sighted by some of them.

"Your first requirement will be a positive mental attitude and a will to succeed. No one can expect to evade capture if your mind is devoid of purpose.

"You're likely to succeed when you know you can and are determined to do so, whatever the odds. The more you train your mind and body before you attempt to evade, the better your chances are for success. The dangerous attitude of 'it couldn't happen to me' should be countered with 'it may happen to me, but if it does I will be prepared.'

"In the first few minutes on the ground in enemy territory you must give quick consideration to landmarks, bearing, and distance to friendly and enemy forces, a likely location for a helicopter landing, or a pickup by hoist and the initial direction to take in your evasion.

"While you're doing that, drink some water. You'll be in some kind of shock from the sudden change in your environment. The water will take the edge off the shock and give you a better presence of mind.

"To be forewarned is to be forearmed. Knowledge of what to expect is important, because when circumstances arise which have been thought out in advance, they can be carried out more quickly and easily.

"If your efforts to escape fail and you become a POW, remember, talking is always the first step to collaborating," Chilton reminded us.

"Once a man starts talking there is no escape from more talking, and the more he talks, the greater his guilt and anxiety. These have very important psychological results. His guilt and anxiety absorb a lot of his psychic energy, which makes him less able to cope with the normal stress and strains of prison life. His whole personality tends to disintegrate. He knows he has done something wrong by submitting to the communist demands and his conscience will not let him rest.

"Beware of this. It's one reason the Code of Conduct was developed. In a nutshell, the code was designed for you to maintain your honor and self-respect while a POW.

"What is your life worth?

"Whatever you think. You must have a cause worth dying for.

"Others before you have. At the Alamo, San Juan Hill, Bunker Hill, or on Iwo Jima. The first tenet of the Code of Conduct is:

"I am prepared to give my life in defense of my country.

"Nathan Hale did when he declared 'I regret that I have but one life to give for my country.' And Colonel Travis was committed at the Battle of the Alamo.

"Next with the Code of Conduct is the phrase 'I will never surrender of my own free will.'

"Third, 'If I am captured I will continue to resist by all means available.'

"Fourth, 'If I become a POW I will keep faith with my fellow prisoners.'

"Fifth, 'When questioned, should I become a POW, I am bound to give only name, rank, service number and date of birth.'

"Sixth, 'I will never forget that I am an American fighting man responsible for my actions and dedicated to the principals which made my country free.'

"Don't believe enemy promises. Remember, Lenin said, 'Promises, like pie crust, are made to be broken.'

"One thing to ease your stress while a POW is to be sure you've properly taken care of your family before going on cruise. Be sure you have an updated will, that your pay records are properly filled out, that your insurance is up-to-date, that your bank accounts are in joint custody, that a power of attorney is left behind. Get the paperwork done to make sure it's easy on your family to fulfill their responsibilities to you while you're away."

Our afternoon was spent with refresher lectures on how to be rescued by helicopter. We were bussed to a Navy dock along San Diego Bay, donned and inflated a Mae West, and were loaded aboard an old World War II landing craft.

Two HUP-2 helicopters from my squadron showed up on schedule. As the landing craft bobbed with the waves at the entrance to San Diego Harbor, the instructors had us jump in the cold Pacific one at a time, pop off a smoke flare, and wait to be rescued. In January the water felt so cold I doubted if icebergs would melt in it.

The approaches by the helicopters were tentative and slow, and while in a hover there was a lot of dancing around. I treaded water and shivered while waiting to be rescued. It seemed to take a long time for each hoist. The helo would lift two of us into the aircraft, one at a time, then climb up to about 100 feet and fly to the beach, land, and discharge two wet and cold survival students. The two helos continued to rotate the pickups until each student had his turn in the sling.

I learned later that this wasn't just training for Survival School students. We were being used as guinea pigs by HU-1 for the final stage of training squadron pilots and helicopter crewmen to practice rescues with live bodies before going aboard ship. It takes a lot of confidence in the training process to accept the potential risk to lives in practicing for the real thing.

During our final night on the beach, I learned a valuable lesson. We had our usual staple—crabs, and boiled them. Everyone ate the crabs this time. As a picky eater, I recognized one important lesson. There are no foods a person doesn't like. It's just that he's not hungry enough to eat some foods. I never again refused to eat something by saying that I didn't like it. I simply say, "I'm not that hungry!"

Early Thursday morning we got a break. We relaxed and slept while we were bussed about 3 hours to the 6000-foot level of the Cuyamaca Mountains and dumped off by the Lost Valley Truck Trail along Highway 79. The instructor divided us into groups of five, and gave us a map, compass, and canteen with water. He also warned us to look out for rattlesnakes, although they might be in hibernation due because of the cold weather. There were two dots on the map. One where we were and the other marked Indian Flats, where we were supposed to be by nightfall. Our parachutes would be at the campsite—about 12 miles away. The incentive was that an AD Skyraider single-engine attack aircraft was scheduled to parachute our dinner at the campsite—some stew.

It was a practice navigation hop for the attack guy in the plane. It seemed to us our life depended on his accuracy of that routine training flight. We also had heard that sometimes the plane failed to show.

The senior officer of each group of five was in command on this hike. I was assigned four enlisted men, including a first-class petty officer from a VP patrol squadron. He was about 38, had a huge pot belly, and had not done strenuous exercise in 15 years. He was in bad shape physically and had no motivation or desire to be in Survival School, either. He provided me a real leadership challenge. The shortest distance between two points is a straight line, and that's how I intended to lead our group.

All five of us were required to arrive at the destination together. The incentive was to be among the first groups to arrive at the destination and receive a sandwich as a reward. A Navy sandwich is two slices of white bread with a piece of meat in between. But in this case, it gave us real motivation to cover the 12 miles of hiking through the brush, and cross valleys, ridges, and mountains quickly. If we had to do it, we might as well be rewarded for it.

Twelve miles didn't seem like a long hike. My petty officer would have preferred to follow the roads no matter how long it took, but the rest of the guys were younger and wanted to be a part of the program. We got ourselves oriented on the map, set out a course for the mountaintops in the distance, and started hiking.

It was easy to begin with; then we got to some heavy brush, but it was still relatively flat. One of the guys got a big rip in his flight suit, but we pressed on.

About halfway along we had a choice. Climb to the top of a mountain peak or walk around it a third of the way from the top. Our group had quite a discussion about whether to go up and over or around the edges. Because of our huffing and puffing petty officer, I elected to skirt around the summit. It seemed like an easier way for him. It wasn't a popular decision, especially since there were a lot of vertical draws that we had to enter and exit. This also increased the distance and the complexity of the ordeal, but it wasn't a straight uphill climb.

As time wore on, nerves frayed and tension grew. The fat petty officer wasn't speaking and preferred just to sit down and wait to be rescued. Several of the guys were willing to let him do it. I was determined that we'd finish the course and we'd do it as a team whether anyone liked it or not.

Then my petty officer complained of a blister. It was a real problem. He hadn't paid attention to common sense about ensuring a proper fit of his socks and boots and failed to keep his shoestrings tight. This allowed his foot to rub in his new flight boot.

I was furious at him for not thinking, but out in the middle of nowhere it didn't make much sense to blow my stack. Our objective was to reach camp—not get into a big fight. The guy who had the ripped flight suit took a strip of the torn material and made a rudimentary bandage as best he could.

Our exhausted, overweight petty officer hobbled on, but our pace slowed. We had to help him over some huge rocks. We pressed on and fortunately didn't get lost.

Although we were in the middle of a California winter, we kept a lookout for rattlesnakes that were rumored to infest the area. The bite threat wasn't as bad as in summer—but we wouldn't get the snake for food, either.

As we closed in on what I calculated was our destination, just before sunset, two ADs swooped over and made a diving pass about 500 feet above the terrain. A small parachute popped open and drifted down about a half a mile ahead. Our morale shot up. The finish was in sight. Our pains eased. Everyone now knew that they could make it.

We aimed for where the parachute appeared to land, figuring it was our destination. When we got there, we found ourselves with a quarter of a mile of rugged bush still ahead of us. The AD navigation had been close, but was not on target. We continued looking forward to hamburger stew boiled in 5-gallon containers. It sounded like gourmet food.

Our group arrived in the middle of the pack. Several of the groups got lost and staggered in after dark, using our bonfire as a navigation aid. It was cold that January night, but our half-parachute sleeping bags combined with our flight jackets kept us reasonably warm.

We all had the next day's activities on our minds—going through escape and evasion and, if caught, 24 hours in a POW compound. By now we were so bone tired, everything in survival situations seemed real and some of the bad things were magnified.

I awoke Friday morning from a fitful night sleep. I was hungry, tired, and uneasy about what lay ahead.

We were given a briefing. The aggressor course was a 2-mile-long canyon to the northwest, from Indian Flats to Freedom Village back near Highway 79. It would be infested with enemy troops. Our job was to evade capture as we tried to weave our way through the brush in the canyon to reach Freedom Village at the other end. We had to do it within two hours, and we were cautioned to stay away from a nearby Prison

Honor camp. Only those who got caught would be considered POWs and thrown into a prison compound and treated as viciously as our POWs had been treated in Korea.

There was no teamwork in this evasion effort. It was every man for himself. The carrot to escape capture would be graham crackers and milk at Freedom Village.

If taken POW, we were encouraged to try escape, but warned if we failed there would be punishment. They didn't give us any encouragement, but if you were lucky enough to break out, it would be because you outfoxed the instructors and guards. The reward for breaking out would be a sandwich.

Then we were off.

Each student scrambled for what he felt was the best track through the course. Some went high along the ridges of the canyon bobbing in and out of the oak shrub. Others were down in the valley where the movement was faster and riskier. Their idea was to make a rapid transit of the 2 miles—to outrun the enemy.

I elected a medium route part way up the side of the hills on the east side. There was plenty of brush to hide in and anyone lower would have to look into the sun to spot me. Foremost I decided to travel slowly. I was careful to avoid twigs and to walk as silently as possible.

About halfway through the course I heard them. The instructors down in the valley shouting and yelling and firing off blanks from their M1 rifles. They were rounding up a group of our students—new POWs.

Also at that time I heard some noise in the bushes ahead.

The enemy was coming in my direction.

I dove under a large scrub oak bush and crawled up under it as far as I could.

In my crouched position the branches were scratching me, my head was buried into my hands, my legs and knees were tucked up under my chin with my rear end sticking out, and I wasn't sure how far. I could only stare at the ground and try to keep my nervous breathing silent.

The footsteps came closer.

I didn't dare move. I was afraid to even breathe.

Then suddenly another student plowed his way into the bush with me.

I didn't say a word.

I didn't dare even move my head to look.

He was panting hard, but did his best to hold his breath.

The enemy troops were near. I could hear them coming our way—crackling the bushes and brush and breaking sticks on the ground with their heavy boots.

No talk, just noisy footsteps.

I stopped breathing. The sweat was starting to drip down my cheeks. The footsteps passed by and faded.

Whoever was in that bush with me couldn't wait. He bolted out.

He made a lot of noise. Voices and footsteps followed. Then gunshots, shouts, and yells. He was captured. I continued to hunker down motionless. I heard the troops march their new POW off to the compound.

They never came back to check the bush for me. I stayed there statue-like in motionless silence for at least a half hour. Several times there were noises and I could hear distant muffled comments from the aggressor troops capturing another student.

Finally I raised my head gradually and twisted around to look. All was silent. No one was in sight. Quietly I walked from the bush to a tree, then from tree to tree. All was silent. Just as I approached Freedom Village, a big cattle triangle gong started clanging. The 2-hour limit was over.

The bad guys had their prisoners, and some of us had avoided capture. We received our graham crackers and milk!

We watched as prisoners were marched to an isolated area near the flagpole surrounded by a circle of painted white rocks. They were stripped naked and shuffled along in their boots because the shoestrings had been removed. It was surprising how much dignity it took away from a man to make him walk naked. He was so much easier to control. The tough students were now embarrassed in their present condition.

Eventually the guards tossed them their flight suits and jackets. They quickly got dressed. Then we were all loaded on a gray schoolbus and driven to a barbed-wire-enclosed prison compound.

Those of us who escaped capture were assigned to watch the treatment meted out to our fellow students who were trapped as POWs. There was no joking or wisecracks. This was really happening. We knew that punishment for any wisecracks by us would be instant imprisonment.

Once inside the compound, the prisoners started to set up a routine. They determined who was senior officer present (SOP). The guards watched and approved of the efforts. But then to add confusion, they entered the compound, grabbed the SOP, and dragged him out.

Outside the compound they stuffed him in a long black box about the size of a coffin. They forced him to lie face down on some bones scattered on the bottom. The guards represented them to be human remains. They jammed closed the hinged lid, pushed the hasp in place, and noisily clipped the padlock closed. All this was in full sight of the prisoners.

The treatment had a twofold purpose. It was to psychologically test the self-discipline of the encased POW, and to look for weaknesses that could intimidate the prisoners. It had to be difficult to breathe pressed against the bones by the lid. This also cut off circulation. Then there was the isolation of the box. Isolation is a real enemy of many men.

One of the main purposes of the box was to try to make the POW realize he had to get his head out of the box, so he could get control of himself and the situation.

One instructor was talking to a guy in a box who was screaming he had claustrophobia and was dying because he couldn't breathe.

"Comrade, if you can't breath, how can you scream so loud?" the instructor asked.

All of a sudden it got very quiet. The POW should have kept screaming—because by being quiet, he was telling the instructor he was gaining control of himself.

A Chinese interrogator once said, "You can control the physical, but not the mental. But if you control the mental, you control the physical."

Claustrophobia was a device used during interrogations to gain valuable information. A second purpose was to try to frighten the remaining prisoners from setting up a chain of command.

The most complex enemy that prisoners faced was fear. Instructors knew it. They knew that fear could paralyze a man into passive acceptance of his fate or could shock him into panic. Either way he became moldable in their hands.

Today those efforts failed.

The captured students had been listening to lectures all week and were trying to psychologically beat the guards. They selected a new leader and they practiced the basic principles of survival. They faced the facts, made a plan, kept busy, and kept trying.

There were a lot of other tricks, both psychological and physical, introduced to make a man talk. Details of the resistance phase of the compound remain classified.

Throughout the afternoon and evening the test of wits continued. The guards would yank each prisoner out, slap him around, and then take him to the dark interrogation room with blinding floodlights aimed in his eyes. It was a tough session to see how far they could mentally push a POW to get a confession, or what could be perceived as a break from discipline of the Code of Conduct.

One technique was to make prisoners stand with his back against the wall, knees bent at right angles until a fellow's legs sagged and collapsed. They would then jerk him up and place him against the wall again until the muscles ached so bad that it was impossible for the man to squat any longer.

All of us watching behind the one-way mirror got a feel of what a true POW camp could be like, watching the suffering of others. We were fascinated by what we saw and thankful it wasn't happening to us.

Meanwhile in the compound, plans were developed for escapes that included organized disobedience against the guards.

Students not incarcerated, but under house arrest, continued to watch all the activities.

Off in the distance I heard the high-pitch prop sound of an AD Skyraider. I looked eastward down the ravine of the canyon and saw the aircraft's wingtips rocking from the bumpy turbulence of the unstable mountain air. He was barely above the valley floor screaming directly for the prison compound. He was on a simulated strafing run. If he'd been firing his 20-mm cannon, it would have torn everything to shreds.

Suddenly the camp guards sprang into action. They herded the POWs into dirt tunnels in the center of the prison compound and hid them under an earthen bunker.

At one end of the bunker was a raised 20-foot platform that I hadn't noticed before. One guard rushed there and manned a 50-caliber machine gun. He aimed it directly at the approaching aircraft and squeezed the trigger. The machine gun rat-a-tat-tat continuously fired blanks and spat out short bursts of flame.

The AD pressed his attack and then thundered overhead. The machine gunner twisted around and followed the aircraft continuously firing as the plane pulled up, climbed out, and did an aileron roll.

The whole sequence took less than a minute, but the troops knew that there was a friendly in the area and they had not been forgotten.

It wasn't a planned event, but the camp instructors had learned to accept the fact that former students on training flights would buzz the camp and give hope and raise the spirits of the beleaguered prisoners—and it worked.

(Five years later, I owned a small Aeronca airplane, about the size of a Piper Cub, and had fun with a low "bombing" run over the compound by dropping 15 loaves of bread. I watched as the machine gun spit flames at me and wondered how many of the loaves fell inside the compound and if any of the prisoners wolfed down any of the nourishment to appease their hungry stomachs. I also dropped several hundred leaflets promising free country music record albums from KSON, the radio station I owned at the time in San Diego, to anyone who escaped from the compound that week. I'm sure the guards were furious. They must have confiscated the leaflets because no one called the station. I just hoped it raised the morale of the troops at the time.

The guards skillfully led several men near the mental breaking point—but never took them over the edge. Those who showed weaknesses under pressure were weeded out. Word was sent to the squadrons of several men who failed Survival School. If they failed, Navy regs said they could not be deployed overseas for combat duty. That meant the end of their careers as aircrewmen or pilots.

The philosophy of Survival School instructors was that if a man couldn't handle the pressure, he'd be a danger to himself and his fellow prisoners in an actual situation.

Throughout the exercise the guards paid close attention to the actions and behavior of each prisoner as he adjusted to life in the compound. They were searching for tell-tale clues and behavior that could be used to intimidate the students when they were under questioning in the interrogation booth.

One guy found a spider crawling on him. He became frightened and in a panic frantically tried to brush the spider off his arm. Finally successful, he smashed it on the ground with his boot. Guards duly made a note.

During his interrogation when they had him bound with a rope in his chair, the POW was belligerent and uncooperative. The guards placed a tarantula on his thigh.

He watched it crawl up toward his hip.

The guy flipped out.

The spider terrified him.

He started yelling and screaming to get the spider off him.

He was helpless to get it off himself.

He spilled his guts to the instructors. He would tell anything they wanted to know, if they only would remove the spider.

They finally removed the spider and washed him out of the program.

I learned 6 months later that he came back and successfully completed the program. After it was over, he stormed back and demanded of his instructors, "Why didn't you sons-of-bitches use the spiders on me again to try and make me talk?"

"You never gave us any reason to think of the idea," they responded.

He had spent 6 months back at his squadron learning to overcome his fear by touching, holding, and handling spiders. He conquered his fear and was bitterly disappointed he didn't get the chance to be tested under adverse conditions to prove to others what he had proved to himself.

The POWs were harassed all night. Virtually no one got any sleep.

But during the dark hours two groups somehow escaped. They remained hidden the required 10 minutes and didn't leave the area of the camp.

They received their sandwiches and an hour of freedom before being tossed back into the barbed-wire compound.

Interrogated by the guards, they refused to say how they got out—hoping others would be successful, too.

Interrogation in the dark room continued around the clock. Weariness from the lack of sleep and food all week took a toll on the POWs. We were forced to watch all night as they struggled with the Code of Conduct.

When midmorning Saturday finally arrived, everyone was exhausted from the exercise—students and instructors.

Chilton presented a debrief of the lessons learned to the class before we departed.

Chilton again reinforced the philosophies he had outlined earlier in the week during the classroom sessions. My admiration for him continued to grow.

Jess McElroy, one of the instructors and a former Korean War POW himself, shared a few parting stories.

"Once a POW, you've got to find out what you can get away with," McElroy said, "and many of you found out this week.

"Those of you who have kids, realize they are the best resistors in the world and set an example of pushing and probing to see what they can get away with," he concluded.

As we filed out and headed for the buses back to civilization, each of us had approximately the same comments for Chilton. "I wouldn't want to go through it again, but I wouldn't have missed it for anything." Or, "This is something I wouldn't do for love or money, but will never forget it for love or money."

I felt cheated, however. I didn't have to suffer through the prison camp and the interrogation process. I'd often wondered how I would have reacted. In July of 1959, a year and a half later, there was no leniency granted to evaders. From then on, the camp staff was beefed up and all students would suffer the compound fate and experience.

I've heard over the years complaints about the way Survival School was run. It was tough; there was some slapping around, and it wasn't an easy experience. There were stories of brutality by instructors, and some made demands to soften the curriculum. There were letters to Congressmen, and in later years things did soften up under the pressure.

But my experience that week was that no one was injured or tortured in such a way that there was permanent damage—our experiences forced us to focus our attention on what could be in store in a real combat situation. The instructors seemed to genuinely care about each individual student, his behavior, his attitude, and how he responded to pressure.

Warner Springs acted as the bully on the block who tried to pull everyone's chain. The school was intended to make students start thinking, especially when you can't hit back.

It was an effort to show that a person, when pressed, has a lot in himself to draw on and can receive a lot of support from the guys with you in a tough situation.

I thought it was the best training I ever had in the Navy.

With the survival situation over, we climbed into the buses with a variety of emotions, depending on the experience we'd been through the last 6 days. But we were all relieved to be finished with the ordeal.

On the way to San Diego we quietly reflected on the past week's experiences. Then we made a stop about 10 miles down the road at a small restaurant in San Ysabel. We tumbled out with visions of gulping down our first decent meal in a week. As we disembarked the bus, we became a rowdy, noisy, and happy crowd, full of energy. I ordered a cheeseburger and chocolate milkshake and gobbled them down with great enthusiasm. Boy, did they taste good!

About 10 minutes later, my shrunken stomach taught me a lesson about cramming it with large amounts of unaccustomed and rich food. I raced out the back door and barfed my guts out. And I wasn't alone.

I guess the bus driver would have been hijacked and forced to stop by our hungry crowd if he hadn't stopped. The restaurant crew was part of the conspiracy about all the food that was being wasted, but they complied with our wishes and orders. Who was going to turn down such an enthusiastic and determined crowd and the money that went with it?

After sleeping on the ground for a week, a soft bed was an adjustment, and much to my surprise, mostly uncomfortable.

The ultimate touch came the following Monday at lunch. My buddy, John Loomis, and I brown-bagged our sandwiches prepared by our wives. As we sat in the ready room to eat and reminisce about the previous week's ordeal, John excitedly dug into his sack for what should have been a tasty lunch.

The pained look that came over his face captured my attention.

Instead of a delicious homemade sandwich, he pulled out of his lunch sack the survival kit rations—a Survival School cereal and meat bar he had taken home as souvenirs.

His wife Nancy had a real sense of humor!

CHAPTER SEVEN

Squadron Life

THE PURPOSE OF all naval aviation is to deploy as part of a battle group of ships so that the president of the United States can project or use naval force as his foreign policy needs dictate.

As a junior Lieutenant jg, I wasn't so interested in the big picture as the president of the United States saw it. I just realized I was starting a squadron training program to prepare for a long cruise. Approximately 6 to 7 months aboard a carrier, and I relished the idea.

My period of buildup and readiness began with a flight schedule that confirmed that I'd been assigned to the HUP-2.

A handful of new pilots, along with second cruise pilots, were selected by higher-squadron officials to fly the HTL/HUL for deployment aboard icebreakers, to support scientific operations at both the North and South Poles. Those were the small bubble canopy helicopters I'd first flown in flight training.

The South Pole operation was known as Deep Freeze III, which was part of the International Geophysical Year. It was an effort to study ecology at the South Pole. Those operations didn't have any potential for combat; they were to support icebreakers, but were complicated and involved dangerous flying conditions. The enemy was extreme weather conditions and whiteouts that created very hazardous flying. Losses averaged one crashed helo per cruise. Several icebreakers with choppers would work in coordination with each other.

Things crews always took with them to the North Pole region was a supply of old, heavy clothing and sheath knives. These were not for themselves, but to trade with Eskimos. These were sought after items in the Arctic. On the other hand, the helicopter pilots sought the plentiful, beautiful, white fox furs possessed by the Eskimos in exchange for the clothing and knives.

For me it was back to basics. I thought I'd learned everything in Pensacola. But the next couple of months in the HUP showed me my basic training had barely scratched the surface for the finesse that was needed to fly helicopters in the fleet.

Precision and smoothness really counted. It wasn't just flying a satisfactory pattern; it was flying an approach, a takeoff, or a hover in an exacting way. At an airfield you could wander all over the place, but aboard a carrier, with 80 other aircraft crammed on the flight deck, space was limited and you had to have absolute control of your helo, especially during adverse operating and weather conditions. Aboard a cruiser it was even more precise, demanding, and complicated because of wind conditions, limited deck space, and the roll and pitch of the ship.

I soon came to realize the most important part of flying a helicopter is a pilot's ability to hover—not just hover, but hover without oscillation, to keep your bearings, direction, and altitude constant regardless of outside influences like changing wind direction, no reference points over the water, gusts, stack gases, lack of horizons, or intimidation by nearby obstacles. To hover properly, you had to know your helo and be able to anticipate each movement it was about to make in response to outside forces and correct for it before the helo forced jerky corrections.

The basic purpose of all Navy helicopter flying revolves around the ability to hover. This includes hover for a rescue or the approach to a landing aboard a ship. Helos exist because of the special ability to hover, and because of that special capability, helicopters are deployed aboard Navy ships.

Our training focused on developing that special skill of controlling the helo in a hover. There is nothing glamorous or prestigious to flying helicopters compared to the white-scarf fighter crowd, no matter how complicated it is to fly. However, once one of those cocky fighter jock's fanny has been pulled out of the briny deep, there's a bonding that lasts a lifetime between the pilots.

A fighter pilot's respect and appreciation for having his hide saved puts the helo pilot at the top of the fighter pilot's appreciation list. A helo pilot becomes an important part of the carrier team by reassuring the jet jocks that they can come back alive if they leave their aircraft for a quick jump into the water or over enemy territory. The confidence that someone cares about them in time of distress helps promote the arrogant and aggressive attitude that has become legend.

Perhaps it is best expressed in this poem that emphasizes the role of a pilot who flys protective cover for a downed aviator.

Jet Jockey's Lament
O Paymaster, pray tell me why,
The prop pilots get the same as I,
Is it not true my plane is faster,
At altitude flight I'm much the master.

Let these peons carry the load,
That's how the hose nose wings got bowed,
I use blinding speed and a loaded gun,
And finish the job in just one run.

I rush to the target, and hurry back,
We jets don't have much use for flak,
I strike the Commie a mighty blow,
Then hasten back to the PIO.

Out of my way you dirty prop,
I'll take the cream, you get the slop,
I'm master of all land and sea,
You get your Charlie after me.

Taxi a plane; How dare you sir!
Give that job to the common cur,
I'm jockey of the mighty jet,
Cross between Superman and Space Cadet.

The time will come O high and mighty,
Your fire'll go out, your planes get flighty,
Down you'll go into the briny drink,
Your streamlined brick is quick to sink.

Please fly my rescap glorious props,
It's true, with me you were always tops,
I need your help, I need it bad,
Take all the press clippings I ever had.

Your friendship I ne'er more will spurn,
I'll save your —, when it comes my turn,
My superiority is a thing of the past,
Please get me the helo, I'm sinking fast!

—*Anonymous*
Courtesy Roland J. Sorensen, LCDR USNR (Ret)

Meanwhile the helicopter pilot is sweating and brooding just keeping his machine in the air and wondering which one of the thousands of moving parts in his contraption is about to fail.

It was back out to the tarmac and continuous instructions in the HUP. Precision approaches, autorotations, and hovering on the spot, including 360-degree turns on a spot, plus tricky continuous downwind hovering.

There was plenty of ground-school training as well with emphasis on safety.

We were reminded that powerlines seem to be a favorite target for all whirlybirds. A one-inch powerline makes a first-class slingshot. But helicopters weren't meant to be fired from a weapon, particularly not backward.

My instructor reminded me that since helicopters land more often per hour of flight than any other type of aircraft, the potential for accidents is greater. The two main types of landing accidents are hard landings following practice autorotations and blade coning during engagement or disengagement.

We were lectured that the HUP has the highest accident rate of any helicopter and pilot factor accounts for 54 percent of all accidents.

Primary mechanical cause of accidents was engine failure.

We learned that although the helicopter may act contrary at times, ditching does not appear to be any more hazardous than with fixed-wing aircraft. In 51 helicopter ditchings between 1955 and 1958, 39 dunkings resulted in minor or no injuries to the occupants.

Five ditchings had one fatality each.

Of the 11 HUPs ditched in deep water, none stayed afloat longer than 60 seconds, and one sank "immediately." Average time on the surface was about 25 seconds. Once in the water, pilots had more luck rolling the HUP in the desired direction for crew escape and holding it there, probably because of the tandem rotor configuration.

It was all sobering information, so sobering that I vowed to myself to pay clear attention to all safety training.

I spent the first 2 months after survival school learning to know the HUP. On one hour-long flight I made 25 landings. We are also required to study the aircraft manuals in detail and learn to know and understand our machine. We had classroom studies to learn about density altitude. It was stressed that failure to understand density altitude and its effect on helicopter operations created a pilot-induced accident category which was second only to landings.

Increased density altitude results in a loss of lift by decreasing available engine power and lowering rotor blade efficiency. Temperature and humidity are two important factors which affect density altitude even at sea level. When at other altitudes the effect becomes compounded and extracts a tremendous operating toll.

Since we deployed in units throughout the western Pacific, we did not have a large maintenance support system. As a result, it was imperative that the pilots and crew members thoroughly understand helicopters—how they functioned and operated.

I flew along with squadron test pilots to learn what they were looking for, including trying to distinguish the various types of vibrations—the normal ones and the dangerous ones. It was a good learning experience. Depending on what was going on with the helo, there were vibrations in your hands, arms, legs, or back or in the seat of your pants. It was important to learn which vibrations were a threat and how to describe them so the mechanics could remedy the problem with the rotor system.

Night flights really got my attention. The HUP had only a basic instrument panel and no artificial horizon. The lights on the ground and the horizon were the only reference points.

At Ream Field we took off toward the dark black ocean at night. It was necessary to cut the daytime pattern short and turn north toward the lights of Coronado and San Diego in order to have a horizon to keep from becoming completely disoriented.

The other night hazard was the fear of the single-engine failing since there had been plenty of failures of the old tank engines operated at high rpms in the HUP. A night autorotation offered little chance for a noninjury landing. It was impossible to find a well-lit landing spot, and at the high sink rate of the HUP, there would be no time to look around or for second chances.

Every little odd-sounding noise at night was magnified in your mind, which made it difficult to relax and enjoy the hop. You were always waiting for some disaster to strike.

Of course, navigation was always a problem, since the only navigation device in this helicopter was a magnetic compass. If you got lost, the most sophisticated direction finding system in the HUP was to land and to ask someone where you were. We were taught to look for landmarks, and your sectional map was clipped to your kneeboard. Since you didn't dare let go of the collective with your left hand for more than a few seconds to grab the cyclic while you scribbled notes with your right hand or read charts, navigation could be difficult when flying alone.

When I was not flying, my duties as PIO were fun, interesting and challenging. As I got better in handling the HUP, there was a natural inclination to want to fly the more challenging flights that became available. The problem was, I had office space in the Administration building, and requests for exotic operational flights such as flying mercy missions to transport injured sailors from a ship at sea to the Naval Hospital in Balboa Park went directly to the Operations Department pilots.

They got all the choice operational flights before any of the rest of us were aware they existed. All the rest of us, assigned other collateral duties, knew was that what we saw posted on the regular flight schedule. We later heard what had been flown from stories in the ready room.

It seems like in every squadron the operations guys always manage to fly the interesting or challenging flights. Things worked okay for me, though. As PIO I learned to deal with people, so I could cut myself out a niche of getting to fly all the VIP and news media flights. I even introduced an Honorary HU-1 Helicopter Pilot card that we issued to all VIP passengers. Word somehow always got back to the skipper, who issued the orders for operations, to assign me to all requests for these VIP and news media flights.

Helicopters were still relatively new operationally. As a result, I tried to make a helo flight an educational experience for my passengers. This included ensuring that the most important of the VIPs sat up front alongside me in the right seat. I'd usually let them handle the cyclic and rudder pedals. Most of them didn't know how to fly an airplane, much less a helicopter, so we naturally danced around all over the sky as my passenger would fight the controls and continuously overcorrect. Generally it was a bumpy ride, but an experience my passenger would never forget.

The poor passenger in the backseat usually was petrified at the antics of his boss. They never said anything, but when I'd glance back, I could see the terror in their eyes. I never let my flying passenger touch the collective, or we'd have had a disaster. Controlling engine or rotor rpm was just too sensitive and critical.

Naturally I would never let my "student pilot" get us in any real trouble. Before that could happen, I'd always nudge the controls or say something reassuring and take over the controls for a minute or two. Usually they'd want to fly only for the first 5 or 10 minutes. After that they would become embarrassed the way they were handling the helicopter and return control to me. But everyone always seemed to appreciate the opportunity.

In the early part of training, navigation was critical. Stories circulated around the squadron of a nudist colony in Harbison Canyon in the eastern part of San Diego County. Naturally it was the destination for the first navigation flight for all newcomers. The goal was to try to find the nudist colony, to suddenly swoop down and grab an eyeful of whatever was displayed on the ground.

My first navigation flight was a disappointment. There was nobody in sight around the camp, and I began to wonder whether the whole idea was just a hoax. But lots of other pilots came back with exciting stories that, I'm sure, were greatly exaggerated. To my knowledge the squadron skipper never got any letters from the nudist colony directly, although I understand some neighbors in the area complained of the thrashing helicopter noise.

A month and a half after finishing survival school, I made my first shipboard landings. With a cruise-experienced instructor in the right seat, we made five landings aboard an LST off Point Loma.

The seas were relatively calm, but there was plenty of concentration needed for the rolling deck and relative winds. The ship cruised along at about 5 knots. The relative wind was off the port bow.

I approached from the starboard side aft of the ship. As I got in close, I brought the helo alongside the ship and transitioned from flying to a relative position on the ship.

In essence, I started flying in formation on the ship.

A crewman on the deck waved me over with his outstretched arms and a flag in one hand to indicate wind direction. I gradually slid the helo over to the landing deck in the center of the LST in a high hover.

Then the crewman on the deck pushed his arms down, signaling for me to land. I gradually pushed down on the collective. All the practice of hovering and using pressures on the control stick instead of movements was paying off.

There was some dancing around. As the wheels touched deck, I bottomed the collective and the wheel strut oleos compressed. We were on the deck and there to stay.

Two crewmen rushed in, slid chocks around the wheels and hooked tiedown straps from the deck to the side of the helo. I held the collective full down, so no gust of wind could catch the rotor system and knock the helo on its side. Taking off was easier. You had to be sure that both tiedowns were free of the helo and pull up smoothly on the collective.

A few days later I made two landings aboard an amphibious ship. The deck on the fantail had more room than the rolling LST, but the ship had more of a fore/aft pitch. In the course of the landing experience, the crew came over to the helo window and invited us to remain for lunch. The ship had just won the award for serving the best chow in the Navy, they bragged. We feasted on steak and lobster before returning to the squadron. Those were to be my only shipboard landings with an instructor pilot before I was to go on cruise aboard a carrier.

Sitting around the squadron ready room, particularly at lunch time, pilots used their brown bag midday break to play a game of acey-ducey or pingpong.

My gauge for judging pilots was that pingpong players made better pilots because their reaction and competitive spirit was honed. The better the pingpong player, the sharper his flying skills. It wasn't 100 percent accurate, but this technique of sizing up strangers worked most of the time.

I was scheduled for my first short cruise in early April. It was to be a 5-day event aboard the *Bennington*. But first I was scheduled to fly for the North Island Survival School hoist training exercise.

On the Monday before the Wednesday event, I spent an hour hovering a helicopter out on the practice pad with several crewmen. We practiced hoisting. It was dangerous work, especially for the crewmen. In those days, crewmen had an unofficial course to qualify as aircrewmen. They didn't receive aircrew wings or anything else to wear on their uniforms, just a notation in their jackets (official Navy record). All that mattered was that they could do the job.

Qualification as a helicopter rescue crewman involved a variety of training courses, including survival school, an extensive course in swimming, and finally, some hands-on experience operating a helicopter hoist. We didn't have a hoist trainer at that time, and the only way to get experience was by actually doing it.

So, as the pilot, I got some experience, too. We hovered out on a pad with one crewman standing on the ground under the hovering helo and one lowering the rescue sling. The rescue sling was weighted with a lead ball about the size of a softball. The weight would assure that the kapoklined sling would float in the water at the right angle for the rescuee to enter it properly. But the lead ball would be dangerous if it hit someone.

The fellow on the ground would get into the sling, while the crewman sitting behind me operated the hoist and lifted our "rescuee" into the helo. He'd then close the trapdoor on the floor, and I'd land. Then they would exchange roles.

The tricky part came after each had practiced several times. The man on the ground would feign incapacitating injury. That required the crewman in the helicopter to attach himself to the hoist cable with a "comealong" called a "Chicago grip"—a device that would clamp onto the quarter-inch cable. A crewman wore the device always. The comealong gripped the cable in a way that the more weight put on it, the tighter the grip.

At this point the crewman took off his helmet, so we had no intercom communication between us. Getting the helmet wet would destroy the

ICS, including the headset and lip mike. He would give me a thumbs-up and I'd flip the toggle switch on the top of the cyclic down with my thumb to lower him to the helpless crewman on the ground. I could talk over the radio by depressing the radio mike switch with my index finger—all the while trying to hover the helo directly over the top of the man on the ground.

Periodically I'd steal a glance away from the horizon and look down the hatch to see how the men were doing. While looking away from the horizon, the helo could dip or wander away from its relative position above the men on the helo pad.

The procedure was for helicopter crewmen to be sure that the injured man had his Mae West inflated, so he wouldn't sink under the waves, and then be sure that he was properly placed and tightly strapped in the rescue sling.

In real life, he had to do all this while swimming and fighting the ocean waves, and get this job done before the numbing cold water would make his fingers useless. He'd generally wear a poopy suit when operating in cold-water areas. That was simply a baggy rubber suit with tight seals at the neck and rubber feet to prevent cold water from freezing him. It was a bulky, cumbersome piece of gear. Just wearing it caused profuse sweating. Once a crewman was satisfied that his victim was safe in the sling, one of two things could happen, depending on conditions at the time.

If a ship were nearby to land on, the crewman in the water would signal me to start the hoist to pull the injured man up to the helo, but not into it. There was a real concern that the rescuee might inadvertently grab one of the controls as he came through the hatch. This could disable the helicopter—of most concern was the engine mixture control. That would shut off fuel to the engine.

Once I had the injured man hoisted, I would rush to the nearby ship, hover over the flight deck, and lower him into the arms of waiting medics. Then I'd rush back to my crewman still bobbing in the water, lower the sling to him, and hoist him aboard the helo.

He was trained not to touch any control switches or levers in the cockpit, so I'd hoist him all the way in. He struggled around, rolled over the floor of the helo, and flicked the switch to close the hatch. I couldn't reach it.

If we were far away from a ship, or hovering on the side of a mountain or over dense trees, the procedure was for me to hoist my crewman back into the helo. He would crawl back into the rear of the helicopter, plug in his headset so that he could talk with me, and then hoist the injured man aboard. It was demanding flying for the pilot because of the long time in a hover and was a very clumsy procedure. But with the limited lifting capabilities of the HUP, there was no way to carry a second crewman to help. So we practiced that several times on Monday.

On Wednesday, after the survival school mass rescue, we tried this procedure in the water with a lifeboat standing by in case there was a problem.

Four of us green pilots and four crewmen scheduled for long cruises, flew two HUPs in formation to the San Diego jetty at North Island—about a 20-minute flight.

We landed on a sandy part of the jetty and rigged our helos for rescue work. To do that, we folded and stored the copilot's right seat by pushing it forward. We hooked up the rescue sling and tested the trapdoor on the floor. One pilot and one crewman of each helo remained on the beach while the others went to work pulling the survival school students from the currents at the entrance to San Diego Bay. Several pleasure yachts slowed down and idled in the channel to watch the operation. I lifted off, headed directly for the Navy landing craft from which the students were jumping into the water, and lighted off smoke flares. It was easy.

The prevailing wind was in the direction from where the students floated in the water. The first rescue wasn't much different from what I had practiced at Ream Field, although I hovered lower. The ground effect from the low hover required less power, but there was some corrosive saltwater spray whipped up by the rotor blades.

The second hoist was uneventful as well. With the four of us in the helo, I pushed off the hover, then picked up translational lift, and we went forward and climbed to about 200 feet. I then turned downwind and flew about a quarter mile back to the beach and landed.

With teeth chattering, shivering, and soaked in their flight suits, the rescuees picked their way barefooted, through the sharp rocks and seashells as they left my helicopter for a nearby bus.

I waited until my squadronmate completed his hoist and then lifted off to pick up an additional two students. With the two helos we set up a rhythm to make the project happen as quickly as possible. After I had picked up my allotted share, the other crew switched places with me while the engine and rotors continued to roar and whirl.

I went to sit in the jetty rocks and watch. Four crewmen had been driven over from the squadron to participate in the crewmen rescue practice session. After all the trainees had been rescued, the landing craft remained bobbing in the channel in case there were any problems with our training session.

We all completed the injured-man hoist experience with no problems, although it took longer hovering, than during rehearsal back at Ream, because of the difficulty the crewmen had moving in the water. Each crewman got one hoist. Flying the helo was more difficult. There were no easy reference points when hovering over the water, which required more concentration, especially with the crewmen in the water.

On returning to the squadron, the helos were thoroughly washed down with freshwater to avoid corrosion. The crewmen all struggled out of their rubber poopy suits and washed these off as well. We were all now fully qualified for shipboard operations, including carrier planeguard.

Sunday afternoon we flew two HUPs over to the *Bennington* docked at the North Island quay wall. We landed on a flight deck partially full of parked fighters with their wings folded. It was a calm sunny afternoon. The crew folded the blades and secured the helos for underway steaming the next morning.

I lugged my flight gear and clothing for the week down to my assigned stateroom.

Below the flight deck a carrier is a maze of passageways, cubicles, and steel ladders with little numbered stainless-steel plaques. The numbering system makes some sense—once you understand it. I never really did understand it, but managed to get myself from stateroom to wardroom (dining hall), hangar deck, flight deck, and Pri-fly. There wasn't much reason to learn my way around the rest of the ship because the familiar areas occupied my time 24 hours a day.

After getting all my stuff settled, I left for home for the evening—excited and anxious about what tomorrow would bring.

CHAPTER EIGHT

Shipboard Operations

THE SHIP WAS scheduled to depart at 0800 hours, and I was aboard by 0700.

Climbing stairs of a platform about four stories high, a person had to cross a gangplank to the quarterdeck, face aft toward the stern, and then traditionally offer a snappy salute to the ship's ensign flying from a staff on the stern. In this case the ensign could not be seen. Then a salute to the OOD is expected with the comment, "Request permission to come aboard ship, Sir."

It's a big event for every Naval officer—seagoing duty as a pilot aboard an aircraft carrier for the first time.

I looked down from the flight deck as the ship's crew cast off the last mooring line and the tugs maneuvered us away from the quay wall and twisted the giant carrier 180 degrees pointing us in the direction to steam out of San Diego Harbor.

It is impressive to stand there and watch the skyline of San Diego fade as we started moving out of the harbor. Squadron plane captains were spotting the jets at the catapult and in designated rows along the edge of the ship. They would be ready when the time came for the nugget jet pilots to practice their shipboard skills.

Our HU-1 crew was preparing the helo for flight ops by unfolding the blades, fueling, and preflighting the aircraft. Leaving the harbor channel and entering the Pacific Ocean, engine turns picked up to increase speed, causing the entire ship to pulsate as the giant propellers bit into the sea.

A stiff cool sea breeze swept across the flight deck. I zipped up my brown leather Navy-issued flight jacket with leather name patch and gold wings sewn on the left chest. I was proud to be separate from the ship's company.

Then it was off to my stateroom to change out of my uniform and into a flight suit. For this week's cruise I'd been assigned one of the large staterooms since there was no air group aboard. Many of the jets would be flying from Miramar out to the ship to bounce and then return to the master jet airbase. Normally junior officers are assigned to bunkrooms way forward, also known as "Boy's Town."

Operating aboard a carrier I learned that the ship's personnel thought we belonged to CAG (commander of the air group); CAG thought we were crazy for flying helos; and the CO and air boss considered us to be a necessary nuisance because of our fuel, weather, and weight restrictions. So the lesson learned was to watch out for yourself and be sure you were the last word in safety for shipboard ops. Nobody else was going to.

Each squadron was assigned a ready room for working, eating, napping, and relaxing.

As helo crews, we had only our staterooms.

Lugging my helmet and Mae West, I went to the wardroom with the officer in charge of our helicopter detachment and had breakfast. Throughout the meal he gave me a briefing on procedures for our flight ops scheduled to commence at 1000 hours.

In the wardroom we reviewed helicopter operations aboard the carrier.

"You'll need to spot the helo in an appropriate position for takeoff on the flight deck and have it preflighted 30 minutes before scheduled launch," he instructed.

"Be sure the crew unlocks the tailwheel before towing it, otherwise the locking pin will break and you'll have a hard time controlling the direction once you engage the rotors.

"Pri-fly will announce 'start the helo' via the 5MC (flight deck bullhorn). Be sure you have a crewman standing by the exhaust with a fire bottle and use the APU [auxiliary power unit] on the first start of the day. A cold engine really puts a strain on the battery," he warned.

"After you get the engine started, turn your radio on to the land/launch frequency. When your engine is running smoothly and temperatures are normal, call Pri-fly for permission to engage the rotors. When you get it, signal to the flight deck director you're ready to engage by circling your right index finger. He will check around and be sure that all blade boots have been removed and no one is going to get his head chopped off when the blades start rotating. He'll give you final clearance to engage rotors. Keep your eyes on him for any directions he might give you and check the flag he'll hold in his hand to signify wind direction.

"Be sure the ship isn't in a turn, heaving, or rolling too much and that winds or gusts of winds don't exceed 20 knots. Once the rotors start rotating, he'll hold up a small red flag until you're to launch.

"When Pri-fly clears you for takeoff, signal to the flight deck director with your thumbs pointing sideways, moving away from each other indicating for him to have the tiedowns holding the helo to the flight deck removed.

"Pri-fly will remove the red flag they are flying and hoist a white flag.

"The crewmen will run forward along each side of the helo to the director so both you and he can see the tiedown straps before you attempt to lift off. There have been too many accidents where only one strap was removed and the helo lifted off prematurely and tumbled over with blades scattering shrapnel in all directions, causing a major accident right on the flight deck.

"When everything is clear the director will give you a green flag, and you are clear to take off into the relative wind. You'll fly out of the chocks.

"A couple of things you want to remember is don't let the ship make any turns while your rotor blades are turning. The gyroscopic effect could roll the helo on its side even if you're tied down. The OPNAV instructions say don't refuel while the engine's going except in an emergency. Fortunately the fueling cap and the hot engine exhausts are on opposite sides of the helo, so if any fuel is sloshed around or spilled from the fueling nozzle, it won't pose as much danger.

"Remember, helo operations are fairly new for everyone on the ship and as a helo pilot it's up to you to ensure a safe operation.

"Those fixed-wing types on the ship really don't understand helos too much, so here's a copy of COMNAVAIRPAC message 240252Z, which gives you authority over your helicopter operations should you be directly and clearly ordered by the ship to do something you feel is unsafe. If anyone gives you any static, just show them this message or cite it over the radio. It's your fanny on the line—so be sure it's crystal clear they're ordering you to violate something you feel is unsafe.

"Let's go topside."

My first flight from the *Bennington* was to be flown with the officer in charge of our detachment as pilot. I was to serve as crewman sitting on the crewman's bench behind the pilot's seat. He gave me a running commentary over the ICS of dangers and procedures.

"Remember as we leave the ship try not to fly over any of the aircraft and never get above 250 feet—that's airplane territory.

"Don't get too alarmed. My first couple of launches I thought this was going to be rescue city. I thought the jets were going to sink into the water. You'll notice as they are catted off, they drop a little just as they leave the flight deck. It's normal until they pick up a little more speed. The F3H Demon is the worst and most underpowered."

We flew around while the ship reversed course to head into the wind.

"Here, I'll fly through the stack gas to show you what to avoid."

The hot engine gases from the ship's boiler bounced us all around, and the fumes had a putrid sulfur smell. I hung on as he had to make exaggerated control movements to try to keep the helo level. But I didn't say anything even though it was scary. It was a graphic lesson of what to avoid.

"Since we never fly above 250 feet during flight ops, I don't need to remind you, but never practice autorotations over water. If there's some screwup you can find yourself and your crew swimming. Save overwater autorotations for the real thing."

"Roger," I responded.

As the ship steadied into the wind, the signalman broke out the Fox-trot (a communications codeword for the letter F) flag on the yardarm, indicating that flight operations were about to begin.

The ship was ready to commence the launch. We moved into position, flying sideways formation along the starboard side of the ship level with the flight deck about 100 yards off the bow. All I could hear was the roar of our engine, my own O-in-C (officer in command)'s comments, and the vibrations of the helo as the first jet was slung off the bow of the carrier. There was no jet engine noise at all, just a few wavy lines from the heat of the exhaust.

After an hour all the aircraft were launched.

"Angel, this is Big Boy. We'll refuel you on the run and then launch you for the recovery."

"Roger," answered my boss.

"I'll share with you a few tips on coming back aboard, then we'll switch seats while they refuel us and you'll fly the next hop with a crewman," he announced over the ICS.

"Incidentally, you're going to find, in spite of what I told you, refueling on the run is common practice aboard a carrier."

"Roger," I responded.

I was eager to get on with flying, but a little sobered by all the activity and danger involved with the carrier flight ops. The time for reflection didn't last long.

"Now when we come back aboard you want to be sure your brakes are on so you don't roll around on the flight deck before the crewmen get the chocks in place against your wheels and your tailwheel is locked," my O-in-C reminded me.

"When the Fox Flag flies from the yardarm, we are cleared to land. We'll come alongside, hover outboard of the deck edge and move over to the designated landing area on the flight deck, hover, and wait for our director to signal us to land. Once we touch down, keep your toes on the brakes, collective firmly down, and wait until you are signaled that the tiedowns are in place. Pri-fly will give you permission to disengage the rotors. Be sure the ship doesn't start a turn downwind while that's taking place.

"If the winds are below 20 knots or the spread between the wind and gust is 10 knots or less, it's okay to disengage the rotors. Twenty knots of wind is max for disengagement and in an emergency, the very max is 30 knots. High winds are tricky. We've had several incidents where winds have pushed the hollow forward blades downward into the fuselage of the helo on disengagement.

"What we're going to do now is refuel on the run so the ship doesn't have to turn downwind for calm winds. We'll end up flying along as we land with the ship's speed with the rotors at full rpm. Soon as you're refueled they'll launch you.

"Normally, as soon as the blades stop, the crew should install rotor blade boots at the tip of each blade and tie them to the helo so—as the ship changes course and the winds pick up—the blades won't flap and snap off."

We got aboard okay, landed, and kept rotors turning at flying speed. The officer in charge locked the collective down, unsnapped his seatbelt and shoulder harness, disconnected his helmet plugs, and moved out of the way, still keeping his hand on the cyclic stick.

I crawled into the pilot seat, wiggled around to get comfortable, strapped in with the lapbelt and shoulder harness, kept the cyclic between my knees, plugged in the radio leads, and looked out the right side window. I could see the redshirts and a fueling hose leading toward the aft end of the helo. The tank inlet where the nozzle entered the helo was out of my range of vision.

Refueling was still taking place. I looked at the fuel gauge in the lower left of my instrument console. It indicated 300 pounds of fuel. I let them put in 400 pounds. At a fuel burn of about 3 pounds a minute, that would give me an hour and a half flying time until the bright red light came on, indicating another half hour until the tanks were completely dry.

We planned the flight to end with 100 pounds left in the tank because the fuel gauge was notoriously unreliable. It supposedly indicated 30 minutes until empty, but sometimes the helo ran out of gas in 15 minutes. This was to cause me a special problem one day soon.

Once strapped in, my crewman seated behind me and fueling completed, I called Pri-fly.

"Big Boy, Angel ready for launch!"

"Angel, cleared for takeoff."

"Ready to go?" I asked my crewman on the ICS.

"Yes, Sir," he responded.

I signaled for the removal of the tiedowns. I hesitated to be sure that both greenshirts ran forward with a tiedown in hand, then received a green flag from the director and slowly lifted off. It was uneventful.

I flew out and took the planeguard position on the starboard stern of the ship in preparation for recovery of the previously launched planes.

Flying sideways took a lot of left rudder and cyclic. We just sat there flying formation on the ship about 25 knots sideways—my first operational hop aboard a carrier. The HUP was rigged to fly sideways without any problems.

The A4Ds, Demons and Skyraiders returned for arrested landings on the *Bennington*. All did their qualifications successfully, a bit ragged on the approaches, but it was a learning process for us all.

The hop lasted 1.4 hours. During dinner in the wardroom that night, I felt euphoria that I had earned my pay.

On Thursday afternoon I was airborne about halfway through the recovery period. Everything was routine. I did my usual instrument scan every few minutes. Cylinder head temperature was normal, fuel pressure normal, oil temperature normal, and forward and aft transmission temperatures and pressures were normal. They were all easy to spot because there was red, yellow, and green tape on the gauge signifying the limits of the needles.

As I scanned around, the needles in the rotor rpm gauge were starting to split. The engine rpm remained exactly where it should, but the rotor rpm was starting to decay—was starting to slow down and get out of the green. It stopped right at the edge of the safe operating area. The needles were no longer married. If the rotor blades went any slower than the needle indicated, the blades would cone up and snap off and we'd uncontrollably plunge into the sea. I checked the rotor jaw engage switch to be sure that the clutch connecting the engine to the rotor system was properly in place.

My mind raced to try to figure out what was happening. Physically the jaws on the flywheel clutch system should make it impossible for the main rotor system to slip and go slower than the engine. But I wasn't sure. Maybe it was just the gauge. The needles began to diverge more. I might be running out of time. I pushed down on the collective to get us closer to the water. If we were going to lose our rotors, I didn't want to do so high in the air. It would be a long and fast fall at 60 knots. Our chances to survive would be nil. We could just drop into the water, or if the forward or rear blades snapped off first, we could do a couple of flips before we hit the water.

I had to trouble the ship with the problem. They still had fuel hungry jets in the pattern. But what choice did I have? If I had a problem, time was critical.

"Big Boy, Angel."

No response. They were busy with the jets. Helicopters don't talk to Pri-fly while aircraft operations were under way except for fuel checks.

"Big Boy, Angel. I have an emergency."

"Go ahead, Angel."

"I need to make an immediate landing."

"Roger, come aboard."

Hugging the water, I slowly moved toward the ship and at the last second raised the collective and flew up to the flight deck level.

The jets in the landing pattern were waved off. The ship was all mine, and every eye was on the helo.

As I slid over the flight deck, I was just hoping everything would hold together long enough for me to land.

I wasn't looking for any of my crew to show me where to land.

I just found the first open area and landed. I waited with my brakes on for the crew to come running toward me with the tiedowns and the chocks to secure the helo.

I kept the engine and rotors turning and motioned for our chief maintenance guy to look at the needles. My right index finger pointed to the problem. With all the noise of the jets on the flight deck and my engine and rotors turning, the crew couldn't hear my shouts—but they saw and understood the problem.

"Angel, Big Boy, what's the problem?" demanded the tower.

There were a dozen jets circling overhead that he didn't know what to do with, and the captain was trying to figure out how long he could stay on the recovery course.

"Big Boy, I've got an indication my rotors and engines are separating and I might have to ditch the helo. I'm going to have to shut down. You need to launch the backup helo."

"Roger," came a somewhat disgusted reply.

The O-in-C came out and take a look. He wasn't sure whether it was a problem or not. The winds across the flight deck were 25 knots. I didn't have any choice but to risk shutdown and hope I didn't bust any blades. The O-in-C raced to get the other helo manned and ready to launch.

The air boss was upset. His whole flight schedule had been disrupted, but he couldn't say too much. The crews spent hours that night poring through maintenance manuals and checking out the helo on the hangar deck.

They finally came to the conclusion the problem was a faulty gauge. No one ever remembered that problem happening in the past. I couldn't figure out what else could have been done to avoid stopping flight ops.

I test-flew the helo early the next morning with a new gauge installed, and everything seemed to work fine. Except for that one incident, everything went smoothly for the 5-day short cruise.

I got in 15 hours of flying, which was a lot because the flight hours were split among the three of us.

CHAPTER NINE

Rescue Payback

BACK IN THE SQUADRON the rest of April saw me flying some overwater hops in the HUP, including radar calibration.

A radar calibration flight required flying tight circles at 1500 feet. After being used to flying in a helo no more than 500 feet above the surface, and usually 10 to 20 feet, 1500 feet seemed very high and scary. The ship calibrating the radars needed a focused target that could hold a position of steady bearing, range, and altitude. On these hops we'd joke about flying so high that we'd get nosebleeds.

I also kept my instrument ticket current flying the SNB and continued to fly the bubble-canopy HTL. At least there were a variety of helos to fly at HU-1.

In late May I was scheduled for another cruise, this time aboard the *Kearsarge*. When you had a cruise scheduled, it always meant plenty of practice flying before deployment. It was an exciting cruise with my first rescue hop on the first day of operations.

Then in early June, Bill Yoakley called. In appreciation for his rescue, North American Aviation was inviting me to be their guest for a supersonic jet ride in the sleek new F100F.

Early Monday morning, June 10th, Yoakley flew a twin-engine Beechcraft Travelaire to Lindbergh Field to pick me up and fly me back to North American Aviation's plant at El Segundo near LAX (the Los Angeles International Airport).

All the brass were there for a VIP tour of the plant that included the X-15 which was being readied for rollout.

The test pilot selected to fly it was Scott Crossfield. He proudly showed me some of the new welding techniques that would allow the skin of the aircraft to heat up to a red glow at hypersonic speeds, yet hold the plane's titanium skin together. The X-15 was destined to fly at 5420 miles an hour. It eventually flew to 354,000 feet altitude.

Crossfield was tall, lanky, and quiet and reserved. The young man had a lot of pressure on him to successfully test the rocketship. Of course his job was the envy of every aviator everywhere. While serving as Civil Aeronautics Board chairman in the early and mid 1980s I got to know

Scott quite well. He served as an aviation technical consultant to the House of Representatives. As I got to know him better there, I found him to be a truly delightful guy with genuine warmth and a fine sense of humor.

Yoakley and I suited up in a special locker room; North American was very generous with their hospitality. They had a bright orange test pilot flight suit all fixed up for me, including nametag plus some of the special automatic Mae West opening devices that Yoakley had talked to me about following his rescue on the *Kearsarge*.

While preparing to go flying, a tall slender guy walked into the room.

"Dan," Yoakley said, "this is George Smith, the guy I told you about who bailed out flying supersonic."

Looking at him, I'd never have guessed what he'd been through.

"Sure glad to meet you," I said, poking out my hand to shake.

"How are you feeling?"

"I'm fine now," he responded.

We continued to dress, and I was fitted out with the G suit. Smith suited up, too. He was going on a test hop himself.

Then we flew the Beechcraft out to North American Aviation's final assembly plant and test facility at Palmdale, north of metropolitan Los Angeles and just south of the famed Edwards Air Force Base test center in the desert.

As we preflighted the silver-polished two-seat F100F Super Sabre, Yoakley explained, "this was first flight-tested five years ago…1953…and except for Chuck Yeager in the X-1, it is the first aircraft to go supersonic in level flight."

All the other aircraft that have broken the sound barrier had to do it in a dive.

"This is a great airplane," Yoakley explained, "and it handles so nice. As a matter of fact, I did the first dead-stick landing in it when I had a flameout a couple of years ago."

"I thought dead-stick landings were forbidden as too dangerous in jets."

"It normally is, but it seemed like the right thing when the engine quit."

The service ceiling of the F100F was 50,000 feet. We climbed the ladder of the cockpit, and Yoakley briefed me on the ejection seat and cockpit instruments.

He was going to accomplish two things with this flight: (1) a complete production test flight on an aircraft that had a few flaws that hopefully had been corrected, and (2) showing North American Aviation's appreciation to a Navy rescue pilot.

I wiggled into the backseat and strapped in, then plugged in the leads to the helmet, put on my helmet, and strapped my oxygen mask over my nose and mouth. I was ready to go. We checked with each other over the intercom, and Yoakley cranked up the engine and immediately taxied out.

There was a full afterburner takeoff with a near-vertical climb. We headed north and continued our climb.

"Let's see how high we can go," Yoakley called back to me on the ICS. Up to 30,000, 35,000, 40,000, 45,000 feet. Our rate of climb slowed. I remembered the pressure chamber training. There was something dangerous about being above 43,000 feet without a pressure suit—if you had a cockpit decompression, but I never said a word. We struggled to 50,100 feet.

"That's about all we can milk out of it," Yoakley called back.

"You ever get a helicopter this high?"

As I looked out at the dark violet sky above me and a slight curvature of the earth on the horizon, all I could do was laugh. It was difficult to laugh against the pressure breathing through my oxygen mask, as I pressed the ICS.

"Let's go down a little and see how fast we can get this stovepipe to go," Yoakley said.

"Here you take it, and let's nose over to about 35,000 feet."

A helicopter required such continuous corrections that it took me a few minutes to develop finesse and get used to the smooth, sensitive jet controls to keep the fighter from porpoising up and down.

I leveled off at 35,000 feet at about 450 knots.

"You want to fly supersonic?" Yoakley asked.

"Sure!"

"Okay. Just push the throttle all the way forward. We'll go into burner, and you fly it. Try to keep it level."

We sliced through the air. It was silent in the cockpit except for a hum and scratching noise in my headset.

"Watch the airspeed indicator," Yoakley said.

"Okay, now you're getting ready to go through the sound barrier, right now. Feel anything?"

It didn't seem any different to me than it did at 250 knots. All I'd heard about bumping and buffeting going through the sound barrier were just stories.

"No, it was just like subsonic flight, although the controls sure are sensitive," I responded as we bobbled up and down a little.

Yoakley was very patient with my flying and my roughness on the controls.

"All right, nose over a little and let's pick up a little more speed," Yoakley instructed.

The Mach meter continued to increase to 1.1, then 1.2 and 1.3 and a little extra needle width. "That's 1.32," Yoakley said. "That'll get us up to 1003.2 miles an hour."

"You ought to have fun with those figures at your squadron."

"Sure will, the HUP couldn't beat it, even in autorotation."

"Let me have it a minute."

He rolled the plane over and did a split S, and we were heading for Mt. Whitney and then at very low altitude, along the Owens Valley and Kearn River Canyon for a close-up look at the High Sierras.

I punched the ICS button.

"Back in high school I climbed Mt. Whitney, and on a bet with my girlfriend at the time, stood on my head on the top. She thought it came to a point at the summit, but in reality it's pretty flat and rocky," I told Yoakley.

He laughed. We climbed back up to about 25,000 feet. "Want to try a few acrobatics?" he asked. I couldn't wait.

I took us through a series of loops, half Cuban eights, barrel rolls, aileron rolls, and all the stuff I'd learned in Pensacola. It was so much easier in a jet than in the SNJ. There was no torque, but a few more G forces involved.

"It's time to go home," Yoakley said.

"Gosh, it seems like we just took off," I replied.

We swooped through Tehachapi Pass and into Palmdale. As we landed, Yoakley popped a drag chute to slow us down and we taxied back to the line.

Yoakley opened the canopy, and the hot desert air felt somehow refreshing. My sweat from the flight evaporated with a cooling effect. He sat in the cockpit a moment filling out test flight paperwork, and a photographer took a lot of souvenir photos of us both.

We jumped in the twin Beech and motored back to North American headquarters at El Segundo. After a shower I dressed in my freshly pressed khaki uniform and went to a ceremony they held with the company president, Ray Rice. Yoakley and some other test pilots also attended.

They presented me with Mach Buster pins, certificates, model aircraft, my orange flight suit, and some stickers and patches. Of course I wore the orange flight suit around the squadron and took a lot of razzing from the squadron for a while. I even acquired the nickname, "Mach-Buster McKinnon."

There was another surprise awaiting my return to HU-1. Crew assignments had been posted for a large group of us junior pilots. I searched for my name on the large, white plastic board listing ships with pilot assignments marked in heavy black grease pencil. There was no individual conference or personal discussion about your assignment—your name was simply posted on the large board.

I couldn't find my name among the detachments scheduled to deploy aboard the carriers, and read through the names again. John Loomis was assigned to the USS *Lexington* (CVA-16). Then I spotted my name: Lieutenant jg McKinnon—Unit 7 USS *Helena* (CA-75).

At first I didn't understand the significance of the assignment.

After asking a few questions around the squadron, I discovered that the *Helena* was to become the new U.S. Seventh Fleet Flagship and was to be home of the top U.S. commander in Asia—the commander of the Seventh Fleet. I'd been selected to be his personal pilot for my long cruise. It would be just me as pilot and four enlisted crew members to maintain the HUP helicopter I'd operate off the fantail of the cruiser.

I never did find out why, as a very junior officer, I was selected for the choice plum of duty, but I started making plans. The first step was to find out who my assigned crew members were. The crew leader was J. D. Dunlop, a gray-haired aviation mechanic first-class. He was an 18-year veteran and had a classic first-class maturity about him. He also liked to dress sharp in his uniform. I learned he could be a little temperamental on occasions, but he was fair with his men.

He was to be assisted by R. S. Wilson, a third-class machinist mate with 7 years' Navy experience and a quiet, shy man.

To handle repairs to the nonengine parts was R. E. Karasinski, metalsmith second-class with 10 years in the Navy.

To handle the radio and electronic repairs was a radioman second-class, D. R. Pobst, the youngest member of our unit.

Parts and supplies were important support for our deployment, especially on a cruiser that would be separated from carriers and bases. This ship was not dedicated to flight operations. I assigned Dunlop acquiring parts and supplies as his first priority. I wanted him to cumshaw (i.e., to obtain outside the normal procurement channels by hook or crook) all the extra supplies that would help support our operations. I didn't want to be AOCP (aircraft out of commission—parts) because the supply system was supposed to have included them on the cruiser and forgot.

I got hold of a metal cruise box the size of a trunk and kept it locked in my office. I rat-packed everything into it I could find, including World War II World Aeronautical Charts of Japan, the eastern coast of China and Taiwan, Hong Kong, the Philippines, the Saigon area, everywhere I expected the commander of the Seventh Fleet to visit. The only cross-country navigation system I had in the helo was a magnetic compass and clock, so the charts would be invaluable if I couldn't get a DF (direction finder) steer from the ship's radar. Basically DF steers were used only over water. So over land I would be back to my days of navigating like it was when I first learned to fly in the Aeronca at college.

I put the wheels in motion to prepare for the cruise, and then had to deploy for another week aboard the *Kearsarge* for more carrier operational experience.

The week-long cruise was uneventful with a couple of exceptions. As we left the harbor the first day, the ship turned northward around Point Loma and took an unusual roll. The squadrons had been moving the planes around to respot the flight deck, and at that moment an A-4 was being towed into position.

The unique roll of the ship, known as the *ballast point sea roll*, caused the A-4 to start sliding with its tug attached. The plane captain smashed the brake pedals in the cockpit with his feet as hard as he could. The tug driver did the same to try to stop the heavy aircraft from skidding toward the edge of the deck and over the side. Their efforts had no effect on the heavy jet.

The crewmen accompanying the plane couldn't get tiedowns attached to it. It continued rolling toward the edge of the deck.

The plane captain and tug driver escaped just as the plane went over the side, taking the tug and towbar with it. It was a freak accident. Divers later recovered the aircraft from the bottom of the sea.

On the second day, late in the afternoon I was flying planeguard. It was supposed to be a short practice launch-and-recovery period for the jets. I launched with the usual 1 hour 30 minutes of fuel.

It was a problem taking off with more fuel, because if the plane on the first cat shot went into the drink, I'd be too heavy to handle the rescue. The launch went off without any major problems. However, it seemed a little slow between cat shots.

"Angel, we've completed the launch. Stand by to recover aircraft."

"Roger," I radioed.

I flew out of planeguard position on the starboard side of the ship into normal flight to get a break from the sideways flying. Then I joined up on the stern of the ship, took my routine planeguard position, and watched as the jets started to make their traps.

I had enough experience to judge the pilots and their skills. For some reason, this was a ragged group. There were lots of bolters. Bolters ate up time. It was getting late in the afternoon. The ship's crew had to be sweating getting everyone aboard.

If there was a real problem, the planes could always "bingo" to Miramar or to an airstrip on isolated San Clemente Island, but that could mess up the schedule of the ship and squadrons. There was great pressure to get everyone aboard and to have made all the necessary qualifying traps.

The ship was getting ready to deploy, and all air group pilots needed to be qualified. This was to lead to a real dilemma for me.

As time wore on I gave Pri-fly the customary call, "Big Boy, Angel, 15 minutes to red light."

"Roger, Angel."

They were well aware of my need to come aboard at red light. They had been briefed on the helo's unreliable fuel system. I listened on the radio. There were still a lot of planes airborne. Then my red light came on.

"Big Boy, Angel, red light on fuel."

"Roger, Angel. It'll be a couple of minutes."

I clicked the mike button twice in acknowledgment, but didn't say anything. As a team player I understood the ship's problem. I had exceeded red light by 5 minutes or so in the past, but I was watching my fuel gauge continuously.

Jets kept coming aboard the ship.

"Big Boy, Angel, 5 minutes on red light," I announced between all the chatter on the radio.

"Roger," came back an annoyed Pri-fly.

The jets had already extended the recovery time. It was starting to get dark, and now the helo driver was adding to the tension. I figured my neck was as valuable as any jet jock.

"Big Boy, Angel, I've been on red light 15 minutes and I could run out of fuel any minute. If someone goes into the water now I don't have enough fuel to rescue them."

"We're trying to get you aboard, Angel. Wait one."

Now I was getting really concerned. My tank had to be almost empty. I didn't know if flying sideways in planeguard position made more or less fuel available to the engine.

"If you don't get me aboard right away, I'm going to declare an emergency," I angrily shot back.

No answer.

There was a jet on final. I kept my silence. He continued down the glideslope and caught a number 2 wire. At that instant the radio came alive.

"Clear to land, Angel," said a relieved voice.

"Roger," I growled.

There was barely enough light to determine horizon, much less see what I was doing as I slid over the flight deck for an immediate landing. My curiosity made me wonder how much fuel was left—burning $^1/_2$ gallon a minute—it would be interesting to measure.

I considered having the crew drain the fuel tank to find out how close I'd come to swimming, but I didn't have the heart. They had to fold the blades, tow it to an elevator, stow it below in the hangar deck, and complete several hours of checks in the dark with flashlights before they could even go to chow. The helo needed to be ready for flight ops first thing in the morning, and aboard a carrier; it's work before food or any kind of play. So there was no use adding to their burden.

But I learned a lesson. Cooperating with the ship by exceeding red light just a little in the past had trained them that the red light wasn't anything to worry about. I never made that mistake again.

There was a scramble getting everything ready for the cruise. We were scheduled to leave the United States aboard the *Helena* on July 16. Most of the pilots assigned as O-in-C of a unit had made a previous long cruise, so they had a better idea of what to take along than I did. Of course, we had the usual inoculations for yellow fever, cholera, smallpox, typhoid, and so on. I naturally got real sick for a couple of days after that.

The plan was to truck our stuff to the ship in Long Beach and pick up our helo from *Det One* in Oppama, Japan, just outside Yokohama and across the bay from Yokosuka, *Helena*'s Far East home port.

My final duty as PIO was staging a flight demonstration at the Coronado July 4th celebration in Glorietta Bay. I asked Lieutenant Hickman to fly the helo, and we worked out a routine based on the tricks of Dumbo, the cartoon elephant. We planned it so that I'd narrate over the PA to 4000 or 5000 people gathered to watch the show. I asked Dumbo to turn left and turn right, dip his nose, and perform other tricks of maneuverability. The helo just seemed to do what was asked of it. It was a good routine, and people seemed to enjoy it.

The squadron got some good feedback. The skipper received a pile of requests for shows from all over the San Diego area. I was glad to be leaving town.

CHAPTER TEN

Long-Cruise Deployment

As WE PREPARED to depart on our long cruise, the fear that another major war was about to explode in the Far East was on everyone's mind.

The Chinese Communists were pounding the small Nationalist Chinese offshore islands of Quemoy and Matsu with a barrage of more than 60,000 artillery shells a day.

The thunder and terror rained from the Chinese mainland across the 6-mile strait on the beleaguered Nationalist Chinese troops. They huddled in underground bunkers on those vulnerable outposts waiting for a respite from the deadly hail of exploding warheads.

It was another confrontation left over from the Chinese Communist victory of mainland China that had forced aging Generalissimo Chiang Kai-Shek and his United States–supported Nationalist Chinese to flee to Formosa in 1949. With each day of increased shelling, worldwide fears were mounting that war would ignite.

The Chinese Communists were on a land grab move again—and the Nationalists were resisting to the best of their ability.

Chiang Kai-Shek was an ally of the United States. The United States couldn't let him go down the drain. At that time, Taiwan was vital to our interests in the Far East as we sought to contain communism.

So everyone knew that the United States was destined to be involved again.

Venomous words were exchanged daily between the Chinese and U.S. government officials.

Just 5 years before, the Chinese Communists had used as many as 1,350,000 of their troops to assist the brutal North Koreans against the U.S. and South Korean forces, in an effort to unify all of the Korean peninsula under communist domination.

Both sides had settled for a truce with no new territory gained. But it caused a loss of 54,246 U.S. lives and an estimated 1,300,000 South Koreans killed and wounded, about one million of them civilians. The Communist Chinese estimated casualties at 900,000 and the North Korean losses at 520,000 killed and injured.

But Communist expansion had been stopped for the moment. Some claimed since they weren't defeated, they would try other areas of advancement.

The Republic of China (Taiwan) territory started at the Red Chinese Beach. There were no equal territorial limits or sharing of the Formosa Strait, nor the islands between the two Chinas. Taiwan was and still is tenacious when it comes to claiming ownership of every inch of the Strait.

Now the Chinese Communists were at it again—in the continuous argument over ownership.

The Red Chinese claimed ownership, but did not have possession of the two islands. These two islands are about one-third the size of New York City, and served as listening posts for the Nationalists. In turn, they served as U.S. intelligence interests.

Some wondered why the islands didn't sink from the weight of the shellings.

Rough seas and the constant shelling of the beefed-up Nationalist forces on the tiny islands made resupply from Taiwan difficult.

Chiang Kai-Shek had said his dream was to return and retake mainland China from the Commies. These islands would be important stepping stones to achieve that goal.

The U.S. government, under President Eisenhower, said it supported the Nationalist Chinese and would face off, once again, with Red China.

In August 1954, Red China's Prime Minister Chou En-lai warned that his government would "liberate" Formosa from the Nationalist Chinese forces under Generalissimo Chiang Kai-Shek. President Dwight D. Eisenhower replied that any invasion of Formosa "would have to run over the Seventh Fleet."

The communist threat then shifted to the string of small Nationalist-held islands—The Tachen, Matsu, and Quemoy groups—off the coast of mainland China. On September 3, 1954, they began shelling the Quemoys and an invasion appeared imminent.

Communist China Premier Chou En-lai had denounced the United States for attempting to create a permanent Chinese Nationalist government on Formosa, in order "to manufacture two Chinas."

Chou said the more the United States "exerts pressure on us, the more we will resist.

"If the United States insists on waging war, we will fight," he warned.

Would the United States go to war, or hold some kind of police action to defend the Nationalist Chinese? Almost everyone in America thought their country would. Would the Chinese Communists think so and back off?

Off and on over the years the shelling continued, but recently the intensity had increased dramatically.

A new flagship for the commander of the U.S. Seventh Fleet was being dispatched from stateside to the Far East. It would be the command post for all U.S. military operations, especially in the Formosa Strait.

During my scheduled 6-month deployment as the commander of the Seventh Fleet's personal helicopter pilot, I'd find out whether the United States intended to go to war or not.

I was to fly from the confined space of the rolling and pitching flagship fantail which barely had enough clearance to allow the helicopter blades to rotate without striking obstacles, including a giant crane and three of the ship's huge 8-inch guns.

Vice Admiral Wallace M. Beakley commanded the most powerful taskforce of U.S. Navy ships ever assembled, which included nuclear weapons and primitive nuclear-tipped Regulus I shipboard surface-to-surface guided missiles.

I kept track of the cruise in my diary.

Wednesday, July 16, 1958

As we were putting to sea from Long Beach, California, it seemed that the war of words was going switch to a war of bullets and U.S. lives.

It was no secret that the entire combined armed forces of the free world had been alerted for the Formosan crisis and Middle East crisis.

The skies were gray and laden with low scuddy clouds. A fresh, brisk, and chilly wind blew. There was spray and mist from the white capped waves.

A call to "set the special sea and anchor detail, make all preparations for getting under way" had been sounded.

The heavy monotone bellow of the ship's deep-throated whistle echoed throughout the busy port. White steam plumed from the ship's whistle mounted on the exhaust stack as the *Helena* prepared to get under way.

It was an order for dungaree-clad sailors who stood on the old wooden piling pier. They scurried around and cast off the last few 3-inch lines binding our powerful gray fighting ship to U.S. soil.

Sailors aboard were wearing starched summer dress whites and stood a couple feet apart as they manned the rails and superstructure of this 658-foot warship. Rows of seamen pulled in unison as a bosun's mate barked orders for the sailors to haul the dock lines through the giant eyehole of the hawsers on the bow and stern of the ship. The heavy ropes would be stored below decks. We were in for 10 days of cruising, heavy training, and drills before we were to leave Hawaii on our way to the Far East.

Eight bells signaled that it was precisely 1600 hours.

Somber, yet full of anticipation, the young sailors in whites and officers in dress blues waved goodbye to loved ones on the pier.

Tears streamed down cheeks from reddened eyes of many of the women as they waved back to their men. Some children playfully dashed around the pier. Younger children were puzzled and wondered why their dads were leaving. With the threat of war, there was a lot of worry as to whether those so precious to them would return at the scheduled time—a long 6 months from now.

Despite the circumstance, this has been the scene of every naval deployment since the time of the Vikings, and is one event that bonds the U.S. Navy from generation to generation.

Hand-painted banners, posters, and white sheets were held by dependents on the crowded pier expressing special thoughts of love or wishing their men farewell.

A small, amateur-sounding, ragtag-looking Navy brass band in white uniforms stood on the pier and played some of the great John Phillip Sousa martial tunes. Without amplification the music didn't sound very powerful or effective. They were a part of the scene instead of dominating it.

The dark gray sky that gave the feeling of sunset, combined with everyone's thoughts about the recent shellings on Quemoy and verbal retorts from U.S. political leaders, cast a dreary mood over the departure.

As we left the pier and pulled into the harbor heading for the Pacific Ocean, everyone aboard forgot about politics and focused on his chores to make our new home shipshape. None of us knew our destiny in the months ahead, but whatever it was, we'd be organized to meet the threat.

The wind whipped up and to make the departing more dismal, light salt spray off the tops of whitecaps misted over the ship. Thoughts of the future and possible dangers involved in the months ahead, swirled around in my head as I watched my wife on the pier fade into the foggy haze.

I walked aft to the small wooden flight deck to analyze the narrow area that would challenge me in the months ahead.

Would I always be able to launch and recover my helicopter safely? It would be a real test of my skills, self-discipline, and character to make safe landings on this postage-stamp-size flight deck.

Our squadron history showed every pilot in recent memory who had operated from the small cruiser flight deck had had some type of accident during his cruise. They weren't necessarily fatal—but always banged up the helicopter or blades.

Then the winds, mixed with the light salty spray topside, became a biting cold as we picked up speed and headed for the open sea. I went below to organize my new living quarters and to meet my new room-mates. I was to be in the middle of the boiling Far East adventure as the heavy cruiser *Helena* deployed to the Far East—an adventure that started for me because of that book (*The Bridges at Toko-Ri*) by James Michener.

Even though there were four officers assigned to my stateroom, it was one of the most spacious. It also had a good location on the ship—farthest aft in officer country. This meant it had the smoothest ride in rough weather. It was also near the wardroom and the supply area.

Junior officers like me were generally assigned up near the fo'c'sle (forecastle) in the very bow of the ship. This area pitches an exaggerated up/down motion in rough seas. Those small, cramped staterooms forward made it tough for a landlubber to get used to the ocean without getting seasick.

I was assigned to my relatively large stateroom because it was closest to the flight deck at the stern of the ship. In case of a person overboard or emergency launch of the helo, I had direct access to all passageways to run to the helo quickly. My new home for the next half year had four bunks, actually two sets of double-deck beds, two desks, a wash basin, and a medicine cabinet. The head was two doors down the passageway.

My crew was berthed in the aft end of the ship, so they could beat me to the chopper to prepare it for a quick launch.

When I found my room, it was dark except for the light of a small fluorescent desk lamp. One of my roommates was already asleep in the sack and it was only 1700 hours—that's 5:00 P.M. I thought, 'What a lazy guy I have for a roommate.'

Then one other roommate showed up. I found out two, Lieutenant Ray Walters and Ensign Joe Cuzzupoli, were in the engineering department; the other, Lieutenant jg Bill Davis, was in communications. My sleeping roommate had the watch at midnight. He was trying to get some rest. Now I knew why they were called "snipes." Because I was second in rank, I got one of the preferred bottom bunks.

A roommate asleep at any time throughout the day during this cruise was to be routine with my engineering roommates. They had watches to stand day and night. They all turned out to be easy to get along with and tolerated my evening procedure of doing an hour or two of paperwork. This task involved keeping flight and maintenance records on the helo, plus reports I had to send back to the squadron.

Each evening at 2150 hours a bugle call, "tatoo," was sounded over the ship's PA system just before taps alerting the crew to "prepare to turn in." That was followed by Chaplain Father K. J. Keaney asking God's blessing on the ship and crew. This happened every night we were under way. Then Taps sounded at 2200 hours and the enlisted crew members were required to turn out the lights.

Thursday, July 17

Hitching a ride aboard the *Helena* was Lieutenant jg Don McCurdy, a squadronmate and his crew, who were being transported to Japan to join Mine Flotilla One aboard an LST in Sasebo. Their command specialized in searching for and destroying mines around harbors. The helo helped in the search for mines and transported a two-star admiral.

Since both of us were just passengers without helicopters on the way to Japan, we spent some time talking with each other.

It was a 5-day trip to Hawaii, and it wasn't particularly pleasant. There were threats of war in the Middle East as well as in the Formosa Strait. Being a passenger on a warship with no responsibility or task to do was boring and actually made me a little bit nervous. I wanted meaningful responsibilities and work that would take up my time and keep my mind off the problems ahead.

We continued drills for fires, general quarters, person overboard, abandon ship, atomic defense, air defense, and theoretical attacks by everything from land-based planes to subs. With the potential for fighting in the Far East, everyone took the drills seriously. If there had been a helo on board, it would have been my battle station. I must be ready to launch for gun spotting or rescue work.

If the ship's big, 8-inch guns opened fire, the helo had to be airborne, or the concussion of the guns' firing would blow out the canopy and probably destroy the rotor blades.

So on this trip to Hawaii I spent most of my time at general quarters as a tourist on the crowded bridge. I watched the action from the nerve center of the ship.

There were two cruisers headed for Hawaii. The other was the USS *Columbus* (CA-74). We cruised in formation along with four destroyers, all heading for the Far East. Lieutenant jg Chuck Fries was the helo driver on the *Columbus*. He brought his helo with him from the squadron.

Our first day at sea he flew over with some brass from his ship. I was on the bridge for Chuck's landing and takeoff. The captain was asking me a dozen questions—is the wind okay, are we heading in the right direction, are stack gasses out of the way, is there too much roll? If I casually mentioned something that ought to be done, orders would be yelled across the bridge. Officers and enlisted crew would run in every direction and would perform whatever I suggested.

I figured the captain, who had only been in command of the *Helena* for 8 months, would be most accommodating for my air operations. While the big confab (confabulation) took place, Chuck and I sat around and traded stories for an hour.

The "blackshoe" Navy is not for me, I became convinced. It's neckties all day long and coats and ties for dinner. On a carrier, you never see a tie the entire time you're out to sea and uniform coats are nonexistent. To keep warm, it's always flight jackets. All this spit and polish, and we don't even have an admiral and his staff on board until we get to Japan.

Sunday, July 20

Disaster has already struck Chuck Fries with his flight ops on the *Columbus*. No deaths or injuries or structural damage, but a problem he'll have to live with. Somehow there was a crew mixup and he engaged his rotors while a blade boot was still attached to one of his rear blades. As the blades started to rotate, the boot, designed to keep the blades from flapping in the wind, actually helped the fragile blade break, and it snapped off.

Each ship carries a set of spare blades. They are 32 feet long. All three come in a matched and balanced set, so if something happens to one blade, all three have to be replaced. The captain of the *Columbus* decided the replacement would take place in Hawaii, not at sea on the windy open flight deck.

I remember the warning from the squadron before leaving CONUS about accidents or incidents effecting every cruiser deployment. Chuck's incident reinforced my effort to be vigilant for my shipboard ops.

Because the helo was broken, the chaplain was highlined over for the Protestant church services. This created a lot of extra work for the ships because they had to steam in formation and toss lines back and forth and set up a little highline cage with a seat to transfer the chaplain.

It was fun watching the flying fish pop up as we steamed along.

The weather became warmer as we approached Hawaii and the ship's air-conditioning didn't work very well, so my engineering roommates rigged four large fans to cool our room.

One benefit to having engineers as roommates is that they know how to take care of the creature comforts.

A chief I'd met on the *Roanoke* during my midshipman cruise recognized me today. He's now assigned to the *Helena*. We traded stories, and I shared how my dream to fly a helo off the fantail of a cruiser began on that midshipman cruise.

Summer whites for dinner tonight, that's about as formal as you get on a ship.

Tuesday, July 22

We pulled into Pearl Harbor, the land of pineapple and poi. It was a beautiful summer day with puffy cumulus clouds. We saw the sunken USS *Arizona* war memorial built to honor those killed in the bombing on December 7th that involved the United States in World War II.

When we arrived at Hawaii the ship had an operational readiness inspection (ORI) from a special unit based on the island of Oahu. The unit goes to sea with the ship for a couple of days and puts the crew through a series of tests, drills, and exercises to determine if they are ready for combat.

There was a lot of importance attached to the inspection, especially since it was graded and reflected on the fitness report of the captain.

The ship would be there for 4 days. I made arrangements for me and my crew to stay on the beach. We were useless to the ship, and I thought I could borrow a helo at Barber's Point. It was also an opportunity to get in some sunbathing, surfing, and sightseeing.

I managed to get myself and my crew a set of orders to NAS Barber's Point at the far end of the island of Oahu. HU-1 had had a detachment there that used to be known as *"Det-Two"* and now was merged with VU-1. All I cared about was that they had a couple of HUPs. It was an hour and 15-minute bus ride.

I talked to the SDO (squadron duty officer) and got a commitment for a helo all day tomorrow, checked into BOQ (bachelor officers' quarters) and took the bus back to town.

Wednesday, July 23

After a quick checkout with the VU-1 pilot, Lieutenant Frank Burgess, I took my crew and started out on a low-level, sight-seeing adventure of Oahu. We buzzed along just above the surf and down to Waikiki Beach, circled around to check the city and girls on hotel rooftops and waves, and then headed out to Diamond Head around to Kaneohe and back down the windward side of the island. What a way to sightsee!

Of course, the real purpose of flying was to get in hours for me and my crew so that we could get our flight skins or flight pay for the month.

I don't know where the term *flight skins* came from, but my best guess is from gambling lingo, referring to "folding money" or "skins" as opposed to "hard money" or coins.

As I reached Camp Kawailoa, my gas started to get really low. I was sweating getting back to Barber's Point. It wouldn't make many friends having to land at some remote point and wait for a fuel truck before getting back to the squadron. Just then I caught a tremendous tailwind and went zooming by Schofield Barracks and back to the base. After a bite to eat, I flew back up to Waikiki Beach for more sight-seeing and returned for some autorotation practice.

I took it easy on the autos. When flying your own plane, you're willing to take more chances than in a borrowed aircraft.

Got in 3.8 hours today, almost enough for the 4.0 I needed to earn my flight pay for this month. I ran into a chief I knew at Miramar who fixed me up with a ride back to Waikiki.

I spent the rest of my time ashore at Fort DeRussy, a military hotel on the beach.

I took a tour of the Dole Pineapple factory. They pack five million cans of pineapple a year. There were plenty of free samples and all the pineapple juice a person could guzzle. It even came out of the drinking fountains. As a big fruit lover, I took more than my fair share. It was great.

Sunday, July 27

We're back out to sea on the way to Japan. Am I ever paying for my pineapple binge of a couple of days ago. That stuff is really corrosive. I was afraid to get more than 20 feet away from the head.

The captain dined with the ship's officers tonight. He usually dines alone in his stateroom. There was a formal candlelight dinner with a ship's band playing soft background music, but, of course, there were no women to enjoy the atmosphere way out here in the middle of the Pacific Ocean.

The back of my legs ache from the blistering tropical sunburn I got laying out on the deck this afternoon.

Tuesday, July 29

The combat drills continue. There were more general quarters alerts. The bosun blows his whistle in a shrill undulating tone on the ship's loud-speaker system. Bells and claxons ring and every crew member races for his battle station.

Everyone wears life jackets, and all those exposed to the outside or enemy fire don gray steel helmets, just like troops in a foxhole. Sound power phones are plugged in all over the ship as a secondary means of communication.

All hatches and doors are dogged (locked tightly) throughout the ship. It's called Condition Zebra. It's impossible to walk anywhere on the ship. Everyone must stay put for hours while communication checks are made and imaginary emergencies are discussed.

We got good news today. We'll hit port in Japan August 4th instead of the 11th. That means one week less floating around with nothing to do. We'll pick up our helos sooner.

The ship will spend 13 days in port at Yokosuka moving the commander of the Seventh Fleet and his staff aboard. HU-1 has a permanent detachment at Oppama airfield where the helos are kept and repaired. It's our squadron's base in the Orient. It's supposed to be just across the harbor from the Yokosuka Naval Base. Me and my crew will spend 2 weeks there.

I'm spending my spare time in the CIC (Combat Information Center) trying to qualify as a CIC watch officer, although they're busy trying to train the ship's regular officers.

CIC is a dark, windowless room in the center of the ship packed with radar screens and data boards. It will be my radio contact area with the ship. From the CIC my radio communications and messages will be relayed to the officer of the deck on the bridge.

Thursday, July 31

Roosevelt may have had his day of infamy at Pearl Harbor, but today was mine. Everyone on the ship had their medical records checked, mine included. Except mine were lost somewhere. "Oh well, what's the big deal," I thought. Except the chief in sick bay said, "No records of your shots, then you take the shots over. Just because you're an officer, you can't get out of it."

"But I already took those back at the squadron and spent 3 days sick in bed at home as a result," I protested.

My appeal to Dr. Guillermo Cabrera, a lieutenant commander, fell on deaf ears. I've heard stories of this happening, but I couldn't believe it was really happening to me. I had to take the yellow fever, cholera, smallpox, typhoid, tetanus shots—the whole darn bunch—all over again. At least so far I feel okay this time.

I spent most of the day trying to send my month-end report message back to the squadron. The captain, XO and Ops officer didn't like the format. I finally convinced them that I didn't create the report, but just filled in the blanks and wrote some essay comments. It wasn't my job to rewrite squadron policy.

One of the big differences between the blackshoe and brownshoe Navy was becoming clear to me. The blackshoes are extremely conservative and just won't take a chance or try anything different. Aviators say, "What the heck, let's give it a try, if it doesn't work so what, at least we tried." I've found sheer will and determination usually make things work. Oh well, as long as one can see the humor in some of these things, I guess it's okay.

The tension between the Red and Nationalist Chinese is increasing. Foreign diplomats have been banned from taking trips more than 12 miles from downtown Peiping (now Beijing) without permission of the Red Chinese Foreign Ministry, and they aren't giving anybody permission. Whatever military movements are going on in the countryside, the Chinese do not want it reported.

Chinese embassy workers in foreign embassies have been arrested, forcing the foreign embassy personnel to do all their own menial chores. To top it off, two patrolling Nationalist Chinese F-86 jets were shot down by Red Chinese MiG-17s in a dogfight near Swatow.

Saturday, August 2

For days we've been steaming along at a fast 20 knots. Now we're ahead of schedule, so today we've slowed down and are steaming along about 4 knots so we don't get to Yokosuka before the 3rd. Instead of waiting til the 4th, we're going into port tomorrow. I can't wait till we get there, get off this ship, and go to *Det One* to get away from all the ship's stuffy regulations.

This morning we refueled from the largest tanker in the Pacific fleet. As they both steamed along on a parallel course, huge black hoses were hauled from the tanker and stuffed into giant fuel ports on the *Helena*. It was tricky work to be sure the ships steered parallel courses. If they got too close together, they could have a collision or cause the hoses to drag in the water, which could pull them out of the tanks and spew warm oil all over the ship and everyone in sight. If the ships steered too far apart, it could jerk the nozzle out of the tank entry or pull the hose apart with the same oil splattering results.

I understood we took on 650,000 gallons of black oil. I wondered how much avgas (aviation gasoline) they took on and how far my car could go on 650,000 gallons of fuel.

Sunday, August 3

We were supposed to arrive in Yokosuka at 0800 hours, but the weather was so foggy that we continuously inched into the harbor with the help of tugs. It was noon before the ship was finally tied up. We were an excited crew to hit port and leave the ship as we loaded our gear on a bus sent by *Det One* to take us on the 30-minute ride to Oppama and our home away from home.

Oppama was basically a U.S. Marine helicopter base, but HU-1 had taken over one of the hangars for office and helicopter workspace. The large *Det One* hangar had a ramp adjacent to Yokosuka Bay. It was situated so that as one took off in a helo, one flew over water avoiding the wall-to-wall and densely populated Japanese homes. Even though it took 30 minutes to drive there along the shore line from the ship, one could see the *Helena* just across the bay from the hangar.

The facilities were obviously Japanese military facilities left over from World War II and now occupied by the United States. After lugging off toolboxes, flight gear, and other supplies, the crew headed for its facilities and I walked over to the BOQ. There was a large comfortable room permanently set aside for visiting HU-1 pilots. There was also a great dining hall with buffet-style serving and a delicious salad bar with plenty of fresh lettuce and tomatoes.

I saw Lieutenant Commander Brown today, my former boss during my brief tenure at Miramar. He's on the admiral's staff and will be moving to the *Helena*. He told me as flag lieutenant he'll be in charge of setting the admiral's schedule and determining when I fly. Since I want to fly a lot, my plans are to keep in close contact with him. When I first met him he was an old jet jockey. Now 2 years later he's trusting me with his life as the right-hand man for the 34th highest ranking admiral in the Navy.

With all the people I keep running into, even as a junior officer, I'm finding how small the Navy really is.

Monday, August 4

Flying again. One of the *Det One* pilots gave me an hour-long orientation ride of the area in the HUP. We flew around like a butterfly to some of the places I'd be flying in the area near Yokosuka. I can't pronounce the funny Japanese names, much less spell them, but it appears the admiral will want to go to some of these helo pads for meetings with Japanese government and U.S. military officials.

Being away from the ship is a relaxed feeling. The ship's crew is busy working 18 hours a day making repairs to the ship and moving the admiral's staff aboard plus all his records and paperwork. I sit at Oppama, trading information on flight ops and learning more about helicopter shipboard operations from squadronmates who have a lot of Far East experience. It's nice because no one at *Det One* tells me what to do. I'm just assigned here TAD (temporary additional duty) as my own boss— really a great opportunity for a young junior-grade lieutenant.

I went to dinner at the home of my old HU-1 XO, who's now stationed at Oppama. Driving there, at dusk, I noticed a lot of short, skinny, worn-down old Japanese men with poles hung over their shoulders with two wooden buckets hanging down from each end of the pole.

"What are those guys carrying?" I asked my host.

"Those are benjo buckets," he responded. "Those old guys are carrying human waste to fertilize the fruits and vegetables they grow in their yards."

"You mean the food I'm eating in the BOQ is grown with that stuff?"

"Yep, but they wash it pretty good before feeding it to you."

I practically gagged and wondered if I could ever eat there again. But then I figured this must be common all over Japan and it looks like everyone has survived.

I'm assigned the newest, cleanest, and best helo in the Far East. It's due to arrive in a few days. It'll keep my crew busy keeping it in top shape, polished, mechanically sound, and free from corrosion.

Flying is nowhere near as dangerous as driving in Japan.

One of the guys at *Det One* took a couple of us "new guys in town" over to Yokosuka in a cab. Yokosuka is the big town nearby and full of shops, bars, and tourist sights. He gave the driver a 1000 yen and told him, "Haiyako."

I wasn't prepared for what happened.

The driver floorboarded the accelerator. The taxi took off like a race car down what was the wrong side of the street for me (left side). Sitting in the right seat I was next to the white centerline of the narrow crowded streets. Our taxi driver, who couldn't speak a word of English, was honking, passing, and forcing the oncoming cars to duck out of the way— before they hit my side of the car. Pedestrians and merchants pushing their carts full of fruit drinks and merchandise scampered to get out of our way. It was like playing chicken, only our driver didn't chicken out.

I was scared silent.

This lasted 15 minutes until we pulled up to the restaurant.

Getting out of the cab, I could hardly stand up because my legs were so weak. The guy with me kept telling the driver, "Dai-ichi bon, dai-ichi bon" which I later learned meant "The best, the best." He had a great time at my expense and could hardly stop laughing.

Later I learned the 1000 yen was a bonus to the driver and the "Haiyako" meant speed up and go as fast as you can to scare the living daylights out of this newcomer to Japan.

Of course it wasn't too many months later that I pulled the same stunt on a new arrival from the States. I sat in the back laughing all the way as he turned as white as I did on my introductory ride.

Wednesday, August 6

The weather in Japan is surprisingly hot, humid, and quite foggy. The headlines in the *Army-Navy Times* today said the Nationalist Chinese government ordered a state of emergency for Formosa, the Pescadores and Quemoy, and Matsu Islands following increased Chinese Communist military activity in the Formosa Strait area. Communist Air Force units have received substantial numbers of new Soviet-built MiG-17 jet fighters and were patrolling the skies along the Formosa Strait intimidating the ROCAF (Republic of China Airforce).

This morning I took off at 0730 hours for shipboard landing practice. I flew a borrowed HUP from *Det One* out to the USS *Jupiter* (AUS-8), an aviation supply ship that transports vital spare parts for aircraft carriers.

The ship had requested a helo for practice flight operations for its crew, and I needed all the experience I could get on small shipboard flight decks, so it made a great match.

The hot and humid weather dramatically affects the density altitude and helicopter operations.

Lieutenant Commander Fortin, the officer in charge of *Det One*, and I flew out with no crewmen aboard. In the hover it took 38 inches of power. I had to nurse the power because with the hot weather we were nearly at max power. This caused touchy approaches which required smooth finesse of the power to be sure the engines didn't overboost or blow.

The *Jupiter* was steaming on the calm waters of Tokyo Bay. The large flight deck mounted on the stern was ideal for practice. Although not as restrictive as a cruiser, it made for good practice. I made seven landings in 2.7 hours.

In the afternoon *Det One* needed testing of the control rigging of a HUP. The correct rigging of a HUP is critical. There is no trim tab to adjust if it isn't properly rigged. This is the one test flight you take slow and easy. The program starts with a run-up on the ground and then a hover. If the controls are within limits, next comes an airborne flight doing all the maneuvers possible to be sure there are no problems at the edge of the operating envelope. Fortunately, everything checked out okay. Another 0.8 flying time, too.

Cameras, pearls, china, and dozens of other items are very inexpensive at the Navy Exchange. Naturally I've been going broke saving money buying everything in sight while the blackshoes are working their fannies off on the ship to get ready for deployment.

A couple of us from *Det One* took a Japanese steam bath and massage tonight at one of the dozens of places available. It's quite an experience. At the front desk, a gal took my money and gave me a towel. I was led to a dressing room and told to strip down. Then a pretty, polite, quiet, and smiling Japanese girl with white shorts and a white bra lead me to a wooden box with a hole in the top. Since neither of us could speak the other's language, she motioned me to walk into the box, turn around, and sit down on a wooden shelf inside.

She closed the doors with my head sticking up through the hole in the top. She then cranked on the steam heat.

Guys were talking about romantic ideas with the masseuses. After 20 minutes in the box, any amorous ideas anyone may have ever thought about disappeared. We were totally destroyed. I mean wilted.

That was followed by a washdown with a hose to get the sweat off so you wouldn't get the water dirty when you got into a hot tub for another 20 minutes. By now I was melted.

Next it was onto the table for a massage. By the time that was over, I barely had energy to stand—but it relieved tension.

Thursday, August 7

I spent the morning doing some flying, sight-seeing, and looking for heliports I would need to ferry VIPs to during the next 6 months. I sure didn't want to get lost with some high-ranking official on board.

With another test flight in the afternoon, I got the problem helo working pretty well. The crew went to the USS *St. Paul* (CA-73), the cruiser we are relieving. I met with Lieutenant Jim Lang's helo crew and tried to get as much helo gear transferred to the *Helena* as possible. We sure don't want to be out of commission for lack of any spare parts.

Friday, August 8

Our state department today charged that Red China had reinforced its airforces in the Strait area to "increase tension and raise the specter of war." All this happened because MiG-17 jet fighters were based at three forward locations on the Fukien coast opposite Formosa.

Saturday, August 9

Today it's official. I'm the commander of the Seventh Fleet's personal helicopter pilot. Lieutenant Jim Lang, his former pilot on the *St. Paul*, left for the States.

Today a State Department policy memorandum was circulated to U.S. embassies that restated the United States' determination not to recognize the Communist Chinese government on grounds that "Communism's rule in China is not permanent and that one day it will pass."

The document added, "By withholding diplomatic recognition from Peiping [Beijing], the United States seeks to hasten its passing."

The memorandum made clear U.S. opposition to the "two Chinas" concept based on recognition of both Communist and Nationalist regimes. It said that President Chiang Kai-Shek's Nationalist government "would not accept any diminution of its sovereignty over China." It asserted that continued nonrecognition would deny Red China access to international councils and make difficult the exercise of its foreign policy and bolster those overseas Chinese and Asians who refuse to accept Peiping domination. The statement also said that nonrecognition was not an "inflexible policy which cannot be altered."

There is still plenty of sight-seeing, but language problems persist. I took a famous on-time Japanese train bound from Oppama to Kamakura— a 10-minute helo hop. I can't speak a word of Japanese, and the train conductors can't speak any English. Finally I figured out I had gone in the opposite direction for 15 minutes. Then it took only 2 hours via train, bus, and foot to get to my destination.

I saw the famous bronze Buddha Daibutsu. It is 44 feet high and is about 700 years old. It's hollow, so I even went inside for a look around. It's a puzzlement to me how anyone can worship a human-made object as a god rather than the God who provided the materials.

On the way back, I picked up a sample rescue patch for the squadron. It's been decided that anyone who makes a rescue gets to wear that patch on his leather jacket right under the squadron patch on the sleeve. There's a street with many stores in Yokosuka known as "Thieves Alley," which specialize in making and selling specialty items for military personnel, including cloth patches.

This afternoon a Marine HRS "Horse" helicopter had control failure and ditched in the water shortly after takeoff from Oppama. I had dinner with the pilot. He was puzzled about what caused his problem, but he felt lucky to be able to ditch in the water.

"What happened?" I asked.

"Dunno! The cyclic just started moving around in my hand like it wasn't attached to anything. I pushed down on the collective. I didn't want to be out of control up in the air. I felt being in the water was the best place to be.

"We hit tail first, busted off the tail rotor, rotated over nose forward, then twisted around and flipped over from the centrifugal force. My copilot and I waited until the motion stopped and the helo started to sink. There were a lot of gyrations. Lucky we had our seat belts on tight or we'd have been thrown all around the cockpit.

"We unstrapped, climbed out the open cockpit window, inflated our Mae Wests, and started to swim. There wasn't time to get a raft out. The water was really cold. Fortunately there was a small Japanese fishing boat nearby that picked us up and brought us to Oppama. They took us to sick bay, but couldn't find anything wrong, so here I am. It was a miracle we got out without a scratch. We're scheduled for another flight tomorrow."

You never give much thought to something similar happening to you, but you do salt away nuggets of information in the back of your mind of what you'd do just in case something similar did happen.

The most important thing I learned from my dinner conversation was to keep my lapbelt and shoulder harness tightly fastened and never try to get out of the helo until all aircraft movement stops.

Sunday, August 10

Today some guys from the *Helena* organized a climb to the top of Mount Fujiyama (Mt. Fuji). Having climbed Mt. Whitney and endured survival school, I didn't feel that it would be much of a challenge. Mount Fujiyama is the highest mountain in Japan.

Only one man from my crew, Don Pobst, wanted to go. I figured the others were too out of shape or lazy. After the difficult climb, I thought they may have heard some inside information about the hike and were a lot smarter than I had thought.

For the last week I had enjoyed the comfort of the air-conditioned BOQ. Last night, I slept aboard the ship to be with the guys going to Mt. Fuji. It was miserably hot and muggy with only a few ventilating ducts.

We left at 0700 hours for the 3½ miles offroad bus ride. We stuffed down a good lunch at a hotel at the base of the 12,390-foot mountain.

We started climbing at 3:30 P.M. During the very first stage I discovered there were stops all along the trail that cater to the tourist hikers. They sell all types of souvenirs, food, and Fuji poles. The Fuji pole was an octagon-shaped long stick with a Japanese rising sun flag attached to the top. The souvenir was used to assist in hiking.

Our goal was to see the Japanese sunrise from the top of the mountain.

Every quarter mile on the steep, rocky, winding trail would be a shack made from corrugated steel and sheets of plywood with sliding glass and ricepaper doors. Some Japanese porters had backpacked food, drinks, and blankets up to each little tourist store.

There were many signs in Japanese, and for payment in Japanese yen, each tourist got his Fuji stick burned with a miniature branding iron with Japanese symbols and letters.

We climbed into the night. I had a flashlight and could make out one of the little huts ahead of the one we just left. The huts with dim candle-lights inside served as a navigation guide along the dark trail.

About 9:00 P.M. we reached a small hut where we were to spend the night. The idea was to get a few hours of shuteye, get up about 3:00 in the morning, and to be on the mountaintop by 5:00 A.M.—in time to see a spectacular sunrise over Tokyo to the east.

Pobst and I had some tea, and I feigned eating sushi—raw fish— offered us by our survival hut innkeepers. I wasn't hungry enough to eat raw fish. We had our Fuji sticks branded, and paid our bill for the "bed," which was the privilege to sleep on a hardwood floor with a blanket. It would have been more comfortable sleeping on the rocky trail.

We hiked up the mountain into low-hanging clouds. We could see only a few feet in front of us all the way to the summit. It was totally clobbered in with clouds and freezing winds. I longed to be aboard the ship with the suffocating heat.

On the mountaintop there were dozens of wooden shacks catering to all the climbers with food, nourishment, and film.

After the miserable effort of hiking to the summit and then seeing nothing but the inside of clouds was a big disappointment. It was impossible to even take a picture.

We huddled around for about an hour with dozens of other hikers, and then headed down the mountain for our bus and the long tiring ride back to the ship.

Boy, did the warm shower and clean clothes feel good. I felt like I'd been to an abbreviated survival school. Climbing Mt. Fuji is one of those things you do only once.

Headlines today included the fact that wartime police force reserves were activated on Formosa. Civilian evacuation plans were issued in case of threatened assaults by mainland Chinese.

Tuesday, August 12

The Chinese Communists launched 90 MiG-17s near Kremoi. It was the first heavy Red air activity since 1954. Nationalist Chinese Air Force pilots engaged the enemy and shot down two Red jets. Red Chinese tactics appear to be the same as we experienced in Korea and the ROCAF is more than prepared.

In order for me and my crewmen to get our monthly flight pay, we had to fly at least 4 hours a month. With our situation unclear and three of my crewmen needing time, I flew 2.1 hours with two crewmen in the backseat. We reconnoitered the area, and I spent some time practicing flying planeguard alongside some of the merchant ships steaming around Tokyo Bay. If I stayed alongside any ship more than a few minutes, the crews and passengers would line the rails and we'd wave back and forth. I'm sure they were wondering what a U.S. Navy helicopter was doing flying alongside their ship.

Friday, August 15

The weather was so foggy today that I couldn't get enough visibility to fly a helicopter until 1630 hours.

I test-flew my assigned helo for the cruise—side number UP-18. It had just arrived on a cruiser on a short cruise, but is basically out of overhaul.

I found a few discrepancies. The crew worked on them and fixed lots of little things on the bird. They're like little boys with a new toy now that they have their own helo. We'll test-fly it again tomorrow morning, but it's basically in excellent shape.

We go aboard the ship Monday morning and then head out to sea. The XO wanted me to fly aboard while the ship was still tied to the pier. The *Helena* is surrounded by carriers, cruisers, and cranes in the shipyard. I didn't think it was a very safe idea because of the cramped quarters, but the pressure increased. If it turns out to be too hazardous, my plan is to fly back to Oppama and then join the ship while it's under way in the harbor. The ship's attitude seems to be "The helo is a real nuisance. Let's get it aboard, tied down, and forget about it as we navigate out of the harbor."

These first couple of hops ought to be hairy without any training for cruiser life. I should learn a lot—fast.

Sunday, August 17

I received news from the States that a good friend of our family was killed in a plane crash. He was Dr. Gordon Dean, the first chairman of the Atomic Energy Commission, a real brilliant fellow and close friend of my dad's. His responsibility was to oversee the development of nuclear weapons in the United States.

The news report said he had scrolled some notes, probably for a speech, on the back of an envelope stuck in an inside coat pocket that was found in the wreckage. I thought they were powerful.

Lessons Learned

1. Never lose your capacity for enthusiasm.
2. Never lose your capacity for indignation.
3. Never judge people, don't type them too quickly, but in a pinch always first assume that a person is good and then at worst is in the gray area between good and bad.
4. If you can't be generous when it's hard, you won't be when it's easy.
5. The greatest builder of confidence is the ability to do something . . . almost anything . . . well.
6. When that confidence comes, then strive for humility; you aren't as good as all that.

7. The way to become truly useful is to seek the best that other brains have to offer. Use them to supplement your own, and give credit to them when they have helped.
8. The greatest tragedies in the world and personal events stem from misunderstandings. Answer: Communicate.

The ship received word that some of our ports of call for the next couple of weeks include Keelung near Taipei on Formosa, Manila and Subic Bay in the Philippines, and Buckner Bay at Okinawa. We're finally getting under way.

Lieutenant Commander Fortin took me aside this afternoon, "This is really the start of your long cruise. Your tour in HU-1 is a rare and priceless opportunity which psychologists call a 'success experience,'" he said.

"Much has been said about other types of opportunities afforded by a long cruise in HU-1: travel to exotic places, exciting adventures, foreign shopping, and all the rest. But little has been said about the success experiences enjoyed by so many of our squadronmates," he said.

"You're about to embark on a life-changing experience. I know, as skipper of *Det One* I'm in the best position to see it happen.

"The typical, newly formed Unit arrives at *Det One* filled with apprehensions. The members fear the unknowns that lie ahead. They are concerned as to whether they can hack the job. They grope along unfamiliar paths. But as the cruise progresses, they start to acquire newly found self-confidence. Confidence born of accomplishments, of being *forced* into positions of responsibility.

"Because they are in a sense a small quasi-independent squadron, they learn to deal with most of the essential aspects of a normal squadron. Their horizons are extended; the scope of their overall awareness is expanded; they learn to see the whole picture instead of a single detail.

"Within a relatively short period of about 6 months, we see formerly timid men miraculously transformed into self-assured dynamic leaders. The transformation is not always pleasant—it is sometimes painful. But almost every alumnus of the HU-1 school is intensely proud of his long-cruise achievements. He has gained an inner power from the knowledge that he has done a difficult job well. To do the job well is traditional in this squadron. Very few have broken the tradition. He faces the future more confidently—be it in or out of the Navy.

"He has enjoyed a 'success experience,'" he continued.

"Contrast our opportunities with those of the usual squadron where an inexperienced man finds himself being third-assistant coffeemaker. At best he becomes familiar with only one little billet. He is required to do everything by the numbers, and somebody else calls the numbers. He never gets the chance to stand on his own two legs.

"Cherish your opportunities, Dan, they may be far more valuable than you can realize at this time. Sayonara," he concluded.

That was powerful stuff. I left inspired. All I could say was "Thank you, Sir," but I never forgot his message and it proved true beyond any possible expectations.

We left the land of cherry blossoms and geisha girls after seeing neither of either so far.

CHAPTER ELEVEN

Shipboard Helo Operations

Monday, August 18

TODAY I ACCOMPLISHED my first cruiser landing without an instructor. I ferried my shiny blue HUP to the *Helena* while the ship was tied up in the repair yards at Yokosuka. I slowly threaded my way in a near hover through other ships and cranes.

No sweat—it worked out okay.

The crewmen brought the last of their parts, equipment, supplies, and personal items to the ship by truck.

We are ready for the long cruise after a lot of preparation.

Once the helo was secure on the stern of the ship, the ship made preparations to get under way. Everyone aboard was anxious to get back out to sea. To be a part of the Navy, hitting port is great, but your work and mission in life is done at sea. We were all restless to get going, regardless of what the future might hold.

As the ship backed out from the pier, we enjoyed a variety of odors only found in a busy harbor.

The ship turned around, with the help of tugs, picked up speed on course, and headed out of the calm harbor waters of Tokyo Bay. It was about then my crew chief J. D. Dunlap sheepishly approached me with my first crisis as O-in-C of the helicopter unit of the flagship.

"Mr. McKinnon, through some mixup, we can't find the special grease for the rotor heads."

"What do you mean, can't find it!" I exploded.

"We've looked everywhere, but somehow we don't have it."

"Ah, nuts. This is no way to start off our cruise. Go back to your locker and do another search. This is a stupid way to start the cruise. The captain of this ship hates helicopters to begin with, and we don't need to create problems."

"You don't think I'd come to you with the problem if I hadn't torn everything apart, do you, Sir?"

"I guess not. Well, what do we do?"

"Could we fly back to Oppama and get a case of the grease?"

"Oh, brother, you want to get me killed?"

Rather than risk having problems with the helo, I went up to the bridge to see the captain and the XO. Needless to say, they hit the overhead, but agreed there was no choice except for me to fly back to Oppama.

Flight quarters were sounded. There was a scramble.

The crane on the stern was turned 180 degrees facing aft over the water instead of over the flight deck. The giant 8-inch guns of the aft turret were fully depressed so they wouldn't interfere with the rotor blades.

The wire mesh railing near the flight deck was pushed out and lowered 90 degrees over the side of the ship so as not to interfere with the launch and recovery of the helo. Hopefully it would also catch anyone who fell over the side of the ship.

Sick bay was alerted.

A lifeboat was manned and swung over the rail so the ship's crew could fish us out of the drink, if necessary. CIC radios and search radars were manned. Fire hoses were strung out on the deck and manned so they could be activated with seawater in case of an accident. The navigator computed the best course with the proper relative winds to keep the stack gases away from the flight deck and still give me enough wind from the correct direction abeam. The required senior OOD (officer of the deck) was already on the bridge because the ship was leaving port. The crew had to respot the helo from a fore/aft position to be lined up cross ship to prepare for the start and launch. This all happened because one crewman forgot his responsibility.

I launched as the ship was outward bound in the channel. The ship continued out to sea as I flew back down the wake toward Oppama. It was a little scary to see both of us going in opposite directions. I hoped something wouldn't go wrong with the helo and the ship would go off and leave me.

I radioed to Oppama. They had my box of cargo on the ramp. When I landed, a crewman opened the door and then shoved a cardboard carton with grease tubes onto the floor of the helo, and I headed off back to the ship.

The *Helena* was picking up steam in Tokyo Bay.

I lined up for my first solo, underway landing. There wasn't much wind. I was a little rough on the controls, and it took a lot of engine boost at slow speed to keep the helo in the air; but I gradually eased the helo over the flight deck. I kept my relative position, pushed the collective down, and nailed it to the flight deck. A little rough, but I was aboard.

While we were going through all the hassle, I figured I ought to get some practice.

"Gladiator, Angel. Request permission to take off for another training landing," I radioed.

"Stand by One," came the reply.

"Oh, boy, wonder what the captain is going to say about this," I thought.

"The bridge says only one more landing, Angel."

"Roger."

I lifted off, nosed over to pick up translational lift, circled around the stern, approached the ship in typical planeguard position, and eased forward.

The cyclic didn't require as much movement as before to keep the helo steady. My second landing was a lot smoother—a real confidence builder.

After the helo was secured, blades folded, and tied down for the night, the crew started refueling.

As the crew put gas in the helo tank, my third-class crewman Wilson discovered the gas from the *Helena*'s tank was contaminated. What a day!

The crew crawled under the helo and drained the tank. The ship's crew didn't want to believe it, but when they saw all the rust flakes and gummy stuff that came out of our tank, they reluctantly agreed that we all had a problem.

After some brainstorming, my crew devised a filtering system. They took a 2-gallon metal bucket and had the ship's company weld a rim around the top. A screened spout was placed to reach from the bottom of the bucket into the helo gas tank. They attached a sagging chamois over the top. It allowed gasoline through, but not the debris and water.

Since there was no way to drain and clean the ship's gasoline fuel tank until we arrived back at Yokosuka, this very slow process became our standard way of refueling the helo.

It was cumbersome, but the chamois screened rust, dirt, crud, and water from the gasoline before it entered the helo's fuel tank.

We could have lost the helicopter, had the contaminated fuel not been noticed. It would have clogged the carburetor shortly after takeoff.

After all this, the crew flushed the helicopter fuel tank several times before we dared to fly it. Luckily there were no flight ops scheduled for the next day.

Pobst, the radio specialist in the crew, wired a talk switch to the side of the fuselage so that passengers could talk to me from their headsets. The regular radio button was on the cyclic stick. But I had a great fear that passengers climbing in and out of the frontseat might trip and hit the cyclic while the rotor blades were engaged. This could cause the helo to go out of control. We had previously removed the collective and pushed the rudder pedals full forward. I didn't want some clumsy passenger messing up my whole day. It wasn't exactly legal to make engineering changes to Navy aircraft without a mile-long trail of paperwork, but we figured that no one in the Pacific was going to object.

At night I pored over maps, and I memorized navigation routes so that I could get the admiral from Keelung to Taipai after we docked in Formosa.

It's really tough trying to read a chart while flying one of these machines. Besides, I had the admiral to impress.

Later I thought about how angry the captain got at me over the grease incident. I thought there was no use ranting and raving like he did, particularly in front of everybody on the bridge. Chewing me out in front of everyone didn't accomplish anything, although it might have made him feel good. The way I figure, what's done is done. I decided to make sure it never happened again.

So I decided not to chew out my crew. They felt bad enough about the incident, and I was convinced they'd do their damn best before they let me down again.

Of course, everyone in the ship's company tries to impress the admiral and worries what he thinks about every little move.

Wednesday, August 20

This was my first day for underway operations. It was exciting to say the least.

We are steaming in formation as a battle group with the aircraft carrier *Lexington* and a ring of sonar-pinging destroyers surrounding both ships for protection from submarines or attack by air.

Admiral Beakley and his staff wanted to have a conference on the carrier concerning the Red Chinese Formosa situation. Flight quarters sounded. With all the staff he wanted to take with him, it took six trips plus one mail round-trip. Our mail is flown to the carrier by COD, then it was my job to get it to the *Helena*. It made me the most popular guy on the ship—when there was mail.

The weather was unsettled and seas rough. Scuddy low clouds pushed by strong winds flew overhead. What made it worse was the wind conditions were very unfavorable for landing and carrying people by helo along the course in which the taskforce was steaming. It required a lot of extra power for me, but drove the engineering crews on the *Helena* almost crazy.

In order for helos to land or take off aboard the *Helena*, with any kind of winds that were marginally safe, the ship had to stop dead in the water. I'd land, pick up more staff, depart for the *Lex*, and the *Helena* captain would order flank speed to get back in formation position with the carrier.

When I returned for more of the admiral's staff, the ship would reverse the four main engines to stop the ship so that I could land.

Down in Main Control bells would ring commands from the bridge signaling the throttle-men to turn valves to control steam to the engines so the shafts would turn or would stop.

I'd load up, take off and the captain would order flank speed again to get back on station with the carrier.

All this stopping and starting is evidently quite hard to do with the giant steam engines that propel this 658-foot ship which weighs 17,000 tons.

Wheels were turning, safety valves were blowing. There was general mass confusion and 123-degree (123°F) heat in the engine room. You could almost hear the engineers swearing through the thick steel bulkheads of their noisy engine rooms.

There was always the pressure on the ship's captain to impress the admiral—which was made all the more difficult because of the demands of the helicopter operations.

While the conference was taking place, I took a break on the *Lex* sharing scuttlebutt with my squadronmates, Lieutenant M. E. Philips, Ensigns B. J. Hale and R. A. Volante, and my best squadron buddy, Lieutenant jg John Loomis.

We sat around for a couple of hours trading stories of flight ops. All of us were starting to feel a little more comfortable with our "at-sea operations." At least that's what we told each other.

We made a trip to the gedunk stand—a mecca for all aboard this giant ship. They treated me to a delicious milkshake. The milkshakes on the *Helena* tasted like they were made with artificial or powdered milk. But on the carrier they were extra good.

"Don't be fooled," John sarcastically told me in response to my excitement about the shakes.

"This ship's not a bad feeder, mind you, but we have noticed that it isn't followed by seagulls, which is a pretty good yardstick."

They also assured me they weren't interested in flying aboard the cruiser, and that I could fly all the trips between our two ships.

Vice Admiral Beakley, better known as "Blackbeard," his radio call sign, was my companion for my eighth cruiser landing. I was naturally a little nervous with him on board, especially with my limited cruiser experience. He was hanging on rather firmly throughout the entire approach. It was a little rough, especially with the lousy conditions, and I didn't blame him for being tense. I got in 2.5 hours' flight time today.

After dinner my engineer roommates would hardly talk to me. Flight operations had totally disrupted their normal routine, and they were legitimately furious. However, before bedtime it became somewhat amusing to sit down and joke about the day's events.

Thursday, August 21

We arrived in Formosan waters. It was blazing hot and muggy. Just before the ship docked in Keelung, I was launched with the admiral to take him 15 miles over to Taipai, the capital of Taiwan. The weather was so hazy that I could hardly see anything. Naturally the ship couldn't maneuver in the harbor and insisted that I take off downwind. The admiral had a schedule to meet, and they didn't want any delays. After studying the conditions, I thought I could make it. So I gave it a try. With the high-density altitude from the heat and downwind conditions, it took maximum power as I nursed the collective up to get us airborne—then the tricky part started.

At 800 feet all I could see was straight down. There was virtually no horizontal visibility because of the smog and haze. I used the map attached to the kneeboard on my right thigh and my jiggling water compass to locate the main airfield in Taipei. I had only a partial instrument panel with turn-and-bank indicator, needle and ball, vertical speed indicator, and airspeed indicator. There were no radio navigation instruments at all, only a two-way UHF radio.

I was about to attempt something voluntarily in a foreign country that I would never have done in the States on a dare.

There wasn't much wind, and I kept my airspeed pegged at 60 knots. At least that way I knew I was going a mile a minute. I'd previously plotted the course and put hash marks at 5-mile intervals on my charts. After the first 5 minutes I got to my checkpoint.

What a relief.

I wasn't there yet, but I wasn't lost yet, either.

The anxiety continued.

I had my helmet face visor down, so the admiral couldn't see my eyes. I didn't want him to know how worried I was.

Finally I reached the boundary of the airport. The only way I could get to the airport was by starting an immediate descent. With tower approval, I started a letdown to the main runway. The smog was so bad I couldn't even see the taxiways the helicopters normally use for an approach. I was nearly caught in an autorotation, because the descent was so rapid.

As we approached the ground it became apparent to the admiral that I'd used the runway for reference. He yelled and screamed above the helicopter noise about following the taxiways. When we finally landed by his honor guard, he leaned over and let me know his thoughts about making approaches to the runways.

He refused to wear a helmet with a headset so he couldn't hear what was going on over the radio. I couldn't talk to him over the intercom, either. As a result, he couldn't hear the tower's approval for my approach and know what was really happening.

I tried to tell him, but he was so busy talking he couldn't hear me. I finally just said, "Yes, Sir, yes, Sir" to everything he said. Then he felt better and got out.

He saw me a little later and smiled and waved to me, so I guessed everything was okay. After the weather cleared a little, I loaded the helo with mail and took off. When you bring mail everyone's nice to you.

Flying over Formosa, I learned that the country is mostly rugged hills and rice patties. The Chinese don't make quite as good use of the land as

Typical planeguard position of the sideways flying HUP as an A-4 Skyhawk approaches to grab a wire aboard the USS *Hancock*.

Rescue sling hits water to discharge static electricity. Rescuee disposes of smoke flare that indicates wind direction and prepares to grab for sling and safety. Pilot is now starting to pull maximum power in the helo as he settles into hover.

Rescuee is hoisted to safety as pilot hovers and crewman operates hoist cable. Once safely in the helo and rescue trapdoor is closed, helo returns to ship or station for medical attention or back to duty for rescuee.

The cruiser flight deck. Railings lowered, 8-inch turret guns are depressed, crane is turned aft, ship is steaming down-wind to provide best wind and roll conditions for helicopter recovery. Note crewman holding green flag indicating a ready deck. There was no radio communication with flight deck or bridge, only with CIC.

The rigging presented plenty of hazards and required steady hovering to avoid disaster during the helo rescue operation of the *Hoi Wong* passengers. Many were so scared on their first ride in any flying machine that they left behind smelly deposits on helo flooring. Some of the ships blackshoes joked that such behavior was to be expected when riding in our early generation helos.

Small craft couldn't reach ship for rescue, leaving our helicopters as the only way to get passengers off the doomed *Hoi Wong*, marooned on reef in the middle of the South China Sea. Up forward, ship's rigging had to be chopped away to provide room for helo to hover for rescue efforts. Waves look deceptively calm. Swells were 10 to 15 feet. It took two helos 4 hours to hoist 106 rescuees to the safety of the *Helena*.

Admiral Frederick Kivette (second from left) personally supervised the arrival and welfare of the rescuees as they arrived onboard his flagship from the *Hoi Wong*. After the successful transfer of all rescuees, messages of congratulations from the CNO, the Secretary of the Navy, and others came to the ship. The story made international headlines showing America's goodwill to China, an adversary at the time.

The ice cream tasted good going down, but turned out to be too rich for the rescuees to keep in their rice diet stomachs. They all barfed it up and made a mess of the ship.

Dressed up in starched test pilot orange flight suit and white scarf for President Chaing Kai-Shek VIP flights. Special plate on the side of the helo of the flag of the Republic of China replaced the three-star plate usually used for COMSEVENTHFLT.

The new HU-1 squadron patch with hand reaching from clouds through angel halo to help distressed hand reaching from the sea was designed by a squadronmate in 1958. The patch was worn by all HU-1 pilots till the squadron was decommissioned 29 April 1994 with a squadron total of 1684 rescues during its 46 year history.

The HU-1 squadron rescue patch worn on flight jackets. Numbers were embroidered in the center of the sling indicating number of rescues made by any squadron pilot. My final patch had 62 in the center.

At the conclusion of an eventful visit and airshow at sea aboard the USS *Midway*, President Chaing Kai-Shek gives a close-up salute to his helo pilot of the day after a safe return to Taiwan. He hammed it up good as I operated my 8-mm movie camera.

Vice Admiral Frederick N. Kivette, CONSEVENTHFLT, says thanks on the fantail of the USS *Helena* following our successful, accident-free cruise. He was my favorite VIP to fly and did a good job of mastering the controls of the HUP on our flights all over the Far East.

Powerful force of water rose over 100 feet in the air through opening of lowered starboard deck edge elevator on USS *Ranger* and totally destroyed a HUP helicopter on the elevator, injuring two of my crewmen and washing two men overboard, one of whom drowned and the other I rescued. The massive giant freak wave was captured on film purely by accident.

One second an angel of mercy, the next the victim of an unforgiving treacherous sea. The destroyed HUP on the USS *Ranger* deck edge elevator after the freak wave hit it.

the Japanese do. The mountains are steeper than in Japan. People are very poor; their clothing is only minimum. I appreciated America even more than I had before I left it.

Friday, August 22

This morning started with excitement. The Red Chinese Air Force was aggressively flying over Quemoy and Matsu, 125 miles to the west, making bombing runs.

All Formosa went on alert.

Our ship scurried to Condition One, just one step below General Quarters or full battle stations.

There was great concern some of the fighters might come over the big island and attack. For a big ship to be tied to a pier in a harbor, with no maneuvering room to fight, is a real nightmare for any captain.

Finally the Chicom (Chinese Communist) planes went away. VIP activities, with all the silver service and niceties, continued as some visiting generals and students of the Formosa Army came aboard the *Helena*.

Saturday, August 23

Today Communist shore batteries intensified their shelling of the Nationalist islands of Quemoy, little Quemoy, Tatan, and Erhtan.

An estimated 50,000 artillery shells rained down on the islands. The previous daily record was 9000 shells fired at the islands June 24, 1947.

The Reds fired so many shells, we wondered how the residents in the area weren't deafened from the noise.

Chinese Communist Foreign Minister Chen Yi addressed the diplomatic reception in Peiping today. It was reported he said, "We are about to liberate" or "We have already begun the liberation of Matsu and Quemoy."

With those reports the tension has started to build.

Secretary of State John Foster Dulles warned today that any Communist Chinese attempt to seize the nationalist offshore islands would constitute "A threat to the peace of the area."

Dulles told the House Foreign Affairs Committee that the United States was disturbed by the evidence of the Communist Chinese buildup, which "suggests that they might be tempted to try to seize Quemoy and Matsu forcibly."

He reiterated that, in the event of an attack on the islands, "President Eisenhower would decide as to the value of certain coastal positions to Formosa."

In other words, Chinese Communists should keep their mitts off Quemoy and Matsu or face U.S. forces.

Dulles reportedly said privately that Chiang had made a "rather foolish" move of committing some 90,000 troops, or about one-third of his effective force, to this highly vulnerable outpost.

American critics of the generalissimo charged that he was trying to suck the United States into an atomic Armageddon; and why die for Quemoy? They also insisted that these beleaguered rocks were not necessary for the defense of Formosa, some 120 miles distant, especially since the U.S. Seventh Fleet stood ready to repel an invasion. But Chiang, who could ill afford to sacrifice so many men, flatly refused to pull back.

Today we pulled out of Keelung Harbor and headed for the Philippines. Our first stop will be Manila Bay. Our plans were to fly to Sangley Point Naval Air Station near Manila, give the helo a good freshwater wash and some land-based maintenance, and return with mail.

We could see corrosion eating away at the aluminum helo from the daily saltwater spray. The ship won't give us enough freshwater as often as we need to wash it down while at sea.

Before departure I had to fly over to Taipei and pick up the admiral. The weather was socked in completely on the way over, so I turned around and followed a river, which was all I could see, back to the ship. The ship was a little disturbed that I didn't pick up the admiral as scheduled. But when you can't see, you can't go. I tried again about 45 minutes later and made it through the haze and smoke, which was actually heavy smog.

I was at the airport 45 minutes early and positioned the helo in the correct location for a full honor guard. The admiral passed through the troops and directly into the door of the helicopter. He was leaving as commander of the Seventh Fleet after our return back to Japan, and most of the countries we visited presented lavish goodbye ceremonies. Beakley had been Commander of the Seventh Fleet for more than 2 years.

At the airport the Chinese military had 200 crack troops lined up for him to inspect. Then he shook hands with the top admirals and generals. While proceeding down the line of troops to the aircraft, a little girl in a starched white dress with bright yellow-and-red trimming presented him with a large bouquet of colorful flowers. As he stood at ramrod-stiff attention before climbing into the helo, the Chinese gave him a 10-gun salute with a cannon that belched blue smoke. Beakley ducked in the helo door, climbed in his seat, and said "Let's go, son." I started the engine, and engaged the rotors, and we blasted off.

The spit-and-polish ceremony caused so many distractions that I forgot to call the tower until I was at the end of the runway, but no one said anything.

On the way we encountered the smoggy weather again. It was so terrible, I was flying on my rudimentary instruments. I thought about flying up to 1500 or 2000 feet to get on top of it, but then figured I'd get lost and never find my way to the ship.

When we finally landed, I got a big smile out of the old man and a big slap on the back. As a Navy pilot he understood my problems, and I'm sure he was as glad to be safely aboard as I was.

The remainder of the day the crew worked on polishing and cleaning the helo.

At night my engineer roommates were busy making preparations for night refueling. During daylight refueling operations, I stood by in the helo to rescue anyone who may fall overboard. Since there was no night flying, I sacked out while the ship's company was busy.

Sunday, August 24

The seas were rough; the strong winds whipped salt spray all over the superstructure of the ship. We were on the edge of Typhoon Flossie. The helo was tightly secured. The crew had so many tiedowns strapped from it and attached to the ship's deck, it looked like a giant spiderweb. We hoped to miss the most dangerous part of the typhoon.

The schedule called for us to enter Manila Bay tomorrow. We needed the freshwater wash more than ever.

Monday, August 25

Today a 10-vessel Communist fleet was routed by Nationalist naval units when it attempted to land troops on Tunting Island, 18 miles southwest of Quemoy. Obvious aim of the PRC (People's Republic of China) troops was to get into position for a further flanking attack. A second Red fleet of five gunboats and 30 motor junks was also driven from the Tunting area later in the day. Two Red ships were sunk, and one Nationalist LST was lost in the action. Communist jet fighters also strafed Quemoy for the first time this year.

The Seventh Fleet, our fleet of U.S. ships scattered throughout the Far East, has been ordered to take "normal precautionary defensive measures" as a result of "the increased activity" in the Formosa Strait.

The Chinese Communist attacks and threats have affected us. As we were entering the channel to Manila Bay, the *Helena* abruptly turned around and headed out to sea. One could *feel* morale drop. There were a lot of long, sad faces. The crew had envisioned a great night on the town in Manila.

Word was we were to head back to Formosa to join up with an assembling U.S. taskforce of warships. We all wondered why we waited until we entered the harbor before heading back up north. My speculation was to get press attention and signal the Chinese Communists that the U.S. Navy meant business.

I took the admiral over to the carrier for strategy conferences and brought back our first mail in a week.

Tuesday, August 26

The Communist Chinese were still striking Quemoy and Matsu. Two of 48 Red Chinese aircraft spotted in the area were shot down by the nationalists. We arrived in southern Formosa late and tied up in the harbor of Kaohsiung on the southwestern tip of the island. Some U.S. forces were stationed at a big Nationalist Chinese Naval facility.

An eight-vessel amphibious force carrying 1600 combat-ready U.S. Marines sailed from Singapore to Okinawa today with a planned stop on the way for combined maneuvers with the Nationalist Chinese on Formosa.

Tomorrow we head for Keelung on the northeast side of the island. Like a normal cruise ship, we do our steaming at night.

Wednesday, August 27

President Eisenhower held a news conference today and drew the line. He said the Nationalist offshore islands were more important to Formosa's defense than they had been 3 years ago. Ike was playing his cards close to his vest when he refused to indicate whether the U.S. military would act to defend the outpost against a Communist invasion, but he said, "The Nationalist Chinese have now deployed about a third of their forces to these islands, and that makes a closer interlocking between the defense systems of the island with Formosa than before."

He added that the United States would not "desert our responsibilities or the statements we have already made." Obviously, that was a warning to the Chinese Communist forces of Mao Tse-tung.

Whatever was about to happen, it looked like I might be right in the middle of the action, even with my operational problems. The carriers in the area continued air ops with fighters armed and bombers loaded with bombs. Swarms of formations of U.S. Navy planes along with those of the Nationalist Chinese were overhead. The Red Chinese were watching all the activity on their radar. They were getting the message that the United States was serious about committing forces to any battle.

We would be having company, too. A buildup was under way. The Pentagon announced the aircraft carrier *Essex* and four destroyers were being shifted from the Sixth Fleet in the Mediterranean to the Seventh Fleet in the Formosa Strait.

We left this afternoon. No one went ashore except for the admiral. The lack of liberty grinded on most of us. We steamed out to sea to rejoin the carrier. I figured we would cruise in circles near Point Delaware—a forlorn spot of nothingness far out to sea—for an indefinite period until this Formosan thing either blew up or blew over.

When I went to pick up the admiral at Taipei Airport, I had to be sure to file a flight plan. The policy was if no flight plan were filed, the guards at the airport would start shooting at your aircraft. About 12 months ago one of the HU-1 pilots forgot to file his flight plan and sure enough they shot at him. The reason was to prevent airplanes leaving that might defect to Red China.

I had a busy afternoon flying 3.2 hours between the cruiser and the carrier transferring people and luggage. I flew planeguard for the carrier for a short while. They were launching F8U *Crusaders* at the time.

There were plane troubles today. On the first part of the hop, the ship's UHF radio wouldn't work and just about the time they got theirs fixed, mine burned out.

The captain was looking for me because the lack of communications caused some difficulties, so I laid low.

The blackshoes are a different breed. It seemed most of us from the brownshoe Navy had to show them that we were different as well. So far I have been wearing my flight suit to chow whenever it appears I might fly. Everyone else on the carrier did the same. I've even worn my baseball cap up on the bridge of the ship, although it's an unwritten "No, no." No one has said anything except some of the guys who wish they could get away with it, too.

Thursday, August 28

This morning I logged 30 minutes' flying time transferring the admiral's staff to the carrier USS *Hancock* (CVA-19). Flight ops were secured prior to noon. Flying was a little tricky because of the high-density altitude. After a good lunch I strolled up to the bridge in my flight suit to kill time watching the ship navigate and talked with the OOD and guys on watch.

We cruised independently, away from the formation of the carrier and its defensive ring of destroyers. It was hot and sunny and the seas were calm, Except for some haze, it was an ideal day.

Then it happened.

"Mayday, Mayday" came a scratchy voice over a loud speaker on the bridge reserved for aviation guard channel—the distress frequency of 243.0.

"Mayday, Mayday" came the radio call again, "this is Rampage 823.

"This is Rampage 823 at 12,000 feet. We're an AD and have had an engine failure. There are three souls on board. Squawking emergency IFF we're...." The voice faded out.

The OOD looked at me. "Sound flight quarters and I'll be ready to take off in 2 minutes," I told him. "You can vector me to him by radio from CIC."

I dashed for the fantail. I guess the OOD checked with the captain first, because I was halfway aft before the bosun's mate gave his undulating whistle signal on the IMC and announced throughout the ship

"Emergency flight quarters, emergency flight quarters. Prepare to launch the helo."

I've never before seen the ship spontaneously swing into action.

The flight quarters lifeboat was prepared to launch. By the time I got to the helo, the ship's crew was depressing the 8-inch guns to clear a path for my rotor blades. Other crew members were ready to rotate the crane aft.

The ship was in a right turn to give me the correct winds.

I climbed into the helo. It was already positioned for flight operations. I donned my Mae West that hung over the pilot seat like a jacket. The crew huddled around, and I explained to them that a plane had had an engine failure and was in the process of ditching.

My crew chief Dunlap appointed himself crewman and Wilson helped him rig and fold the right front VIP passenger seat forward so that he could get the rescue hatch open. I strapped myself in with the seatbelt and shoulder harness, put on my helmet, and started the engine.

Karasinki and Pobst were untying and yanking the rotor boots off the tip of each rotor blade so they could engage.

The crew got the seat rigged and folded forward in record time, and Wilson scooted out of the helo closing the door.

Dunlap grabbed the rescue sling from the side panel and snapped it on the overhead hoist cable right below the lead ball. The equipment was ready, and then he strapped himself to the seat behind me.

"Gladiator, Angel. Ready to engage."

"Angel. We have about 30 seconds more before we straighten out on course."

"Roger."

The extra tiedowns had been removed. All that was left were the standard two, one on each side holding the helo to the deck, plus the chocks on the wheels so that we wouldn't roll forward or backward.

Karasinki and Wilson were manning the tiedowns. Pobst was standing by waiting for any signs from me.

Warrant Officer Mesler was standing about 20 feet in front of the helo near the edge of the ship waiting for me to signal him by rotating my right finger in a circle. This meant I was ready to engage the rotors.

He checked around to be sure the rotors were clear of any obstructions and all hands were clear from the danger zone.

"Angel, Gladiator. We're settling on course, clear to engage."

"Engaging," I radioed back, simultaneously signaling to Mesler.

He gave me an "All clear." I took my right gloved index finger and lifted the safety cover of the engaging switch and pushed the toggle switch forward.

The clutch engaged the rotor blades.

They rotated slowly, but gradually picked up speed. Finally they were at flying speed.

I did a quick mag check, put my right hand between my knees with thumb extended, and moved it back and forth signaling Mesler that I was ready to launch. He ordered the crew to remove the tiedowns. They did.

He signaled me that the helo was no longer attached to the ship.

Up collective and I was airborne.

It had been less than 4 minutes since I left the bridge.

"Angel, your course is 150 magnetic about 12 miles."

"Roger, Gladiator."

I climbed up to 200 feet and added enough power to cruise at 75 knots—maximum cruise speed.

My mind wandered. I didn't know what to expect. I said a silent prayer that no one was injured. I figured the Skyraider could ditch successfully. The seas were ideal for recovery, but could the crew all get out of the aircraft before it sank, or would someone get trapped inside?

Survival school and swimming had taught me that the sea has enormous power and is very unforgiving.

I didn't know what ship these guys were from, but I knew I was going to do my best to rescue whomever was out there. The carrier helo had to be on its way, too. I wondered where they were.

"You ready, Dunlap?"

"Yes, Sir."

"Be sure you have a smoke flare ready to mark the spot in case we need it for some reason," I ordered.

"Yes, Sir," he answered through his intercom.

He had unstrapped his seatbelt and was kneeling on the helo floor looking ahead through the forward canopy helping me locate the downed aircraft.

"Angel, Gladiator, you're 5 miles out from where the IFF disappeared on our scope."

"Roger, Gladiator. We don't see anything yet."

We ground out the distance. We were going so fast the blades were buffeting and I was near a blade stall. I backed off on the collective a little.

Then I saw a green dye marker in the water up ahead. In the center were three one-man yellow rubber rafts attached to each other.

I slowed for a downwind leg and turned upwind—all the time dissipating airspeed.

One of the men lit a smoke flare to give me proper wind direction for my rescue approach. I had slowed down to 45 knots.

"Open the hatch," I told Dunlap.

He opened it and lowered the sling halfway down the hatch. I came to a hover in front of the nearest raft. Three men were visible.

"Gladiator, Angel. Three men and three rafts. Approaching for the rescue," I radioed.

I was too busy flying to respond to any more radio messages from the ship. They had the essential information.

One of the men in one raft pointed to the other two men indicating for me to rescue them first. I figured he must be the pilot and the other two were his crewmen.

We moved in for the rescue.

Dunlap lowered a sling into the water, and we dragged it forward to discharge static electricity. To make things easier, one crewman jumped in the water and put his arms correctly through the horse collar so that it held him around his back. He signaled he was ready to be hoisted.

We got him into the helo. He crawled dripping wet with a slight shiver to the back of the helo to leave room for Dunlap and so that there would be room for his squadronmate to be brought into the helo.

Then Dunlap lowered the sling for the second man. His weight came on the helo. The large range of the helo center of gravity was a big help. I needed nearly full collective power or about 40 inches to hover and hoist the added weight of the second man aboard the helo.

Maximum power is 42.5 inches. Beyond that I'd overboost the engine.

The high density altitude was robbing me of power. Four adult men is about the maximum load of the HUP. I thought to myself, "Should I risk trying to hoist the third man into the helo and maybe overboost the engine, resulting in us all being dumped into the drink, or let the carrier helo I saw approaching off in the distance rescue him?"

If I was successful, I had three rescues here for a total of four. I'd beat out the guys from the carrier rescuing any men from their own ship.

Then I thought, as I moved off the cushion of the hover to gain translational lift to get flying, that I'd need more than 42.5 inches of power.

I realized that getting back aboard the carrier with deck wind wouldn't be a problem, but hoisting the third man into the helo might be.

My calculations continued. Maybe I'd better leave the pilot behind. No use being greedy. He appears to be in good shape aboard the one-man raft.

Just then the carrier's HUP came within about 100 yards of me.

"We'll let him have one," I thought.

Dunlap had hoisted number 2 aboard. He was sitting dripping wet on the floor with a big grin on his face.

"Close the hatch," I told Dunlap. "We'll leave the other crewman for the other helo."

I lifted the collective, nudged the cyclic forward, and moved off the hover cushion and picked up translational lift. We orbited around the rafts.

It took max power, but no overboost. I had made the right decision.

The other helo came in for the third pickup.

After the rescue was completed, we flew in formation toward the carrier, leaving the three tied together rafts behind. I asked Dunlap to have our rescuees write their names on my kneeboard. I radioed back to the *Helena* that we had rescued ATN Donald Rhodes and AD2 Jack Stroud.

Both men were armed with standard-issue .38 caliber pistols.

A tradition in the rescue business was that the rescuee gave the rescuer his personal weapon as a token of appreciation. The gun can be claimed "lost at sea" and a new weapon requisitioned without cost to the downed airman.

One of the soaked and shivering hands reached forward, tapped me on the shoulder, and held forth a .38 revolver by the barrel with the handle pointing toward me. I thought it would be great to have another revolver, but I already had one of my own. And I imagined those two kids would need one for their next flight. So I smiled and waved my head "No." They looked disappointed as I turned down their goodwill gesture.

It wouldn't be many days before I thought it might have been wise to accept the gift.

Once the other helo pilot made his pickup, the *Helena* gave me a heading for the carrier. I switched the radio knob to Channel 7, the carrier frequency, and led the formation. It took about 15 minutes for us to reach the *Hancock*.

Flight ops on the carrier had come to a standstill. Both helos were waved in for simultaneous landings. The yellowshirts and whiteshirts with red crosses swarmed around my helo to retrieve my two soggy catches. Some of the carrier helo pilots ambled up to my window with envious looks of disgust because a cruiser pilot had horned in and made the pickups of some bird farm residents.

It wasn't long before I received a copy of the rescue reports from the crewmen. One read as follows:

STATEMENT OF STR0UD, J.W., 478-67-28, AD2(CA), USN

Approximately 1330, 28 August 1958, AD-5N Buno. 132610, was launched from USS HANCOCK (CVA-19). The crew of this aircraft consisted of Carl E. OATES, LTJG, USN, 596077, Pilot; Donald W. RHODES, ATN, USN, 3742377; and Jack W. STROUD, AD2(CA), USN, 4786728. RHODES was flying left hand rear, Radarman and I was flying right hand front, ECM operator.

This particular flight consisted of two AD-5N aircraft Modex 822, BuNo. 132605, and Modex 823, Buno. 132610. H.E. THIBAULT, Piloting 822, was the flight leader and C.E. OATES, piloting 823, was flying wingman.

When the two aircraft had rendezvoused, course of 350 degrees magnetic was flown and we began climbing to the assigned altitude of 20,000 feet. Approximately 19 minutes after takeoff we had reached 12,000 feet of altitude when the engine of BuNo. 132610 emitted a number of loud, sharp back-fires and the aircraft began to shake violently. Immediately Mr. OATES notified the flight leader that he had a rough running engine and was going to return to the ship.

The TACAN indicated that HANCOCK was bearing 180 degree magnetic, 32 miles from our position. Mr. OATES executed a left hand turn to a heading of 180 degrees magnetic while in this turn Mr. OATES called to my attention that the oil pressure had dropped to zero (0).

The engine was running intermittently and all attempts to regain engine power failed.

"Mayday" and a position report were transmitted to the ship on the UHF radio and Mr. OATES told me to select the "emergency" position of the IFF. Evidently the IFF wasn't transmitting because the ship requested that we "squawk" emergency. The red button that is incorporated in the emergency portion of the IFF would not stay engaged so I depressed it manually and maintained it in that position until we hit the water.

Mr. OATES notified the flight leader and HANCOCK that he didn't think it was possible to make it back to the ship. After all attempts to regain engine power failed Mr. OATES turned the battery switch to the off position.

The external gas tank was jettisoned at approximately 9,000 feet. RHODES and I were given the order to prepare to "ditch." Mr. OATES took command of the situation and instructed RHODES and myself as to the procedure for "ditching."

All loose gear was stowed in the rear to keep it from flying forward on impact.

Both front seats were lowered and all shoulder harnesses were locked and tight.

At 2,000 feet RHODES suggested that the rear canopy be released. Mr. OATES gave permission and the rear canopy was jettisoned. When the rear canopy had cleared the aircraft the front canopies were opened.

The gradual descent was continued and the next thing I remember was the tail of the aircraft hitting the water. The tail hit the water first and then the forward part of the fuselage hit with a very hard impact.

All three of us were out of the aircraft and in the water immediately after the forward motion of the aircraft had ceased.

The aircraft stayed afloat no more than two (2) minutes and we were clear of the aircraft as it began to sink.

It took approximately five (5) minutes for all three of us to get in our life rafts. Between five (5) and ten (10) minutes had elapsed before we had joined and tied our rafts together. When the helicopter was in sight Mr. OATES released his smoke-flare.

We were picked up in the order designated by Mr. OATES, I was first, RHODES was next, and then Mr. OATES.

Twenty (20) minutes after entering the water we had been picked up by the helicopters from the USS HANCOCK AND USS HELENA.

I believe the major factor in the outcome of this emergency situation was due to the calm and collected reactions displayed and the orders and instructions issued by Mr. OATES.

I have been a designated Combat Aircrewman since December 1957, and have approximately 225 flight hours. My primary rate is AD2.

Jack W. Stroud

Communications were lousy during the rescue. When I got aboard the ship, the captain called me to the bridge and asked, "Did you get my message?"

I hadn't.

He said, "I radioed you not to overload your helo."

I was polite, but what I thought was, "Here I am, always fighting for safety of flight operations even when the ship doesn't give me the proper winds. Now he radios me how to fly while I'm out by myself making rescues I've spent a lot of time training and preparing to do."

I was thankful I hadn't received the message over the radio, although we both had thought the same thing.

The carrier had a big celebration and cake ceremony for the rescue pilot. Since I was aboard the *Helena*, I had nothing. A rescue to the black-shoes on a cruiser means nothing. Aviators make a big deal out of it.

It's been 2 weeks with no U.S. mail—better known as the "U.S. snail." The crew anxiously awaited my every landing from the carrier with hope that I had mail for them. My guess was that stateside letters were being flown from Japan or somewhere out to one of the carriers and mail would reach us soon. The old admiral was ready to hit the overhead because of this predicament.

The XO finished off my day. He ordered me not to wear my flight gear to chow in the wardroom. His contention was that on the flagship, he wanted only uniforms visible. I usually fly until 1915. He expected me to shower and dress in the proper uniform for dinner—dinner which ended at 1830 hours. It didn't make sense to me.

Friday, August 29

Mail call today! I hauled three bags—about 200 pounds total—from the carrier. It really didn't seem like much for the 1200 men on this ship.

There were a lot of staff transfers between the carrier and cruiser, including Admiral Ramsey, COMCARDIV. He was quite a leader, always friendly, always talked, and made me feel important. He sure contrasted with Admiral Beakley, who usually had a glum and sour disposition. However, I continued to have a lot of respect for Beakley.

Several Marine colonel pilots were on one of the trips from the carrier. One of them wanted to fly. So as soon as we lifted off the carrier, I gave him the rudder pedals and cyclic. He couldn't get the thing turned in the direction of the *Helena*, and we gyrated around a bit. His buddy in the backseat was about to barf from the rough flying. Finally the carrier called me and informed me what I already knew—I was headed the wrong direction. I had to take over.

The XO had never been in a helo before. I tried to get him to fly with me so that he could better understand and appreciate my operational problems. I think he was a bit chicken.

I managed to bum a few newspapers off the carrier today. News was 3 days old, but still news to us. The carrier receives *Stars and Stripes*, the military newspaper, which is like a stateside daily. It was really the only way military people had to read news in the Far East.

Psychological warfare has started.

Peiping radio, the mouthpiece of the Red Chinese, usually had a girl disk jockey we've named "Peking Polly." She even dedicated songs to us. She said we'd soon be in Kobe, Japan. Her basic job is to play with our minds and to try and destroy our morale. Most of us just laugh at her comments. Today she had a man on the broadcast with threats that the Nationalist Chinese garrison on Quemoy was a "hopeless place" because of all the communist sea, air, and artillery pressure. They also said a Red Chinese "landing on Quemoy is imminent."

The broadcast went on to say that the Communists were "determined to liberate Taiwan [Formosa] as well as the offshore islands," meaning Quemoy and Matsu. The radio broadcast issued an ultimatum to the Quemoy troops to "stop resistance immediately and return to the fatherland or be totally destroyed."

Our State Department answered the broadcast saying that the Communists were confirming the American view that "the offshore islands are intimately related to Taiwan."

The U.S. government warned that "it would be highly hazardous" for the Communists to assume an attempt to "change the situation in the Formosa Strait and hope it "could be a limited operation." In other words, such action would lead to all-out war.

The U.S. Defense Department said they felt the Communist bombardment of Quemoy and massing of forces on the mainland nearby were considered "serious enough to warrant concern," but regarded it primarily as a "political bombardment." Nevertheless our forces were making a lot of movements.

Besides the aircraft carrier *Essex* and four destroyers being shifted to the Seventh Fleet from the Sixth Fleet in the Mediterranean, the carrier USS *Midway* (CVA-43) was also being shifted from Hawaii to the Seventh Fleet. This action raised the strength of our naval force near Formosa to six carriers, two cruisers, 36 destroyers, four submarines, and over 20 support and amphibious vessels. The U.S. Air Force deployed to Formosa with a squadron of jet fighters and some transport aircraft.

Today the Communists bombarded the Tan Islands. These islands are composed of a 96-acre island and a 40-acre island, $2^1/2$ miles south of the Amoy Port and 10 miles west of Quemoy. They also belong to the Nationalists, but hug the Communist shoreline. Tung Ting, another 40-acre island, was 17 miles south of Quemoy, also was bombarded. It was also disclosed that the Communists have set up a group of MiG-17 jet fighters on a new airstrip at Shati—just 9 minutes' flying time from the Pescadores and about 12 minutes from Taiwan.

Sooner or later with all this talk and rhetoric, things have to come to a showdown.

Saturday, August 30

With the threat of war heating up, this afternoon the admiral impetuously wanted to know how long it took the ship to fire one of its 8-inch guns.

He found out.

From the moment he gave the order, it was a matter of minutes until the blast shook the ship and nearly broke hundreds of unsuspecting eardrums. I was in my stateroom and rushed topside to find out what happened.

The gun used in the test was on the turret farthest aft—right near the helo.

The concussion from the blast tore the superstructure off the chief of staff's barge. It was a 50-foot boat usually lowered over the side to take the admiral's number 2 man from ship to docks when the *Helena* was in port. The hull was cracked and the barge was not operational.

It was a very expensive test.

Standard procedure was to fly the helo off the ship before firing the big guns. The shock and blast may blow out the plastic canopy in the nose of the helo. Structural damage is the next concern.

We miraculously found no visible damage to the helo, but I wanted to test-fly it before I took the admiral on any flights.

The whole ship buzzed because of this incident.

Sunday, August 31

I test-flew the helo at 0600 hours. I was a little apprehensive after the big blast which caused the destruction of the chief of staff's barge. I couldn't believe the helo had withstood the shock. But everything seemed to handle okay. I first hovered over the flight deck for about 5 minutes to see if anything would pop or break. When everything held together, I took off and circled around the ship, and stayed close enough to be picked up if something broke or quit. I was careful not to gain too much altitude.

Soon I ferried the admiral to the carrier. None of the brass were laughing about the destruction of the chief of staff's barge, although there were plenty of jokes among the crew about the lack of foresight that caused the incident.

With very rough seas and high winds, we learned that Typhoon Gracie was somewhere in the area. But the destroyers in our formation wanted church services and requested that the chaplain be delivered to them. Someone decided to send him via helo. This was to be my greatest challenge to date.

The tin cans were bouncing around like corks. I came over the fantail and set up a hover. The stern of the destroyer bounced up and down at least 20 feet. I had to fight the controls to maintain my position. This took a great deal of cooperation between me and my crewman. He talked to me over the intercom, giving me directions to come forward or move back or to the left or right, to maintain my exact position over a clear space of the deck on the fantail of the destroyer and away from the depth charge racks.

It was so rough that I couldn't steal any glances downward as I normally did. All my attention and concentration were focused on trying to fly a formation or a fixed position in relation to the towers and stacks of the destroyer.

Since the ship was bobbing up and down in such extremes, I elected to hover as close to the ship as possible, so there would be less slack in the cable as we lowered the chaplain. As the chaplain climbed into the rescue sling wearing his Mae West, in preparation for lowering, I wondered if he had given any thought to just how dumb this stunt really was. He just looked over at me with a big ear-to-ear grin with his white teeth shining. His expression was one that made me think he thought he was going on a great Disneyland ride. I hoped his prayers were current.

The deck crew all had kapok life preservers on, and each man was fastened by a tether cord to a railing or some piece of equipment on the deck. Their feet were stretched out wide in a bracing stance.

White water was foaming, spraying, and splashing over the bow. The canopy was full of salt spray, making it hard to see.

We finally got the chaplain started down toward the deck. The ship's crew held out their arms and grabbed for him. Wilson stopped the hoist for a moment and then suggested we should try to get the helo in a modified up/down oscillation with the bobbing of the ship's fantail. That way, if we had too much line out while the ship was down, it wouldn't slam up and smash the chaplain.

I couldn't get in sync and was having a heck of a time keeping my relative position. Miraculously we got him down. Wilson had plenty of slack in the cable, so the chaplain wouldn't get jerked around while getting clear of the sling. When he was completely out of the hoist cable and the sling, I lifted up the collective for quick altitude to get rid of the slack and to keep the cable away from the ship so it couldn't get tangled up or wrapped around anything. Then Wilson hoisted the sling into the helo.

We gradually improved our hoist operation with each ship we visited, and by the end of the experience, it dawned on the chaplain just how dangerous this procedure was. We transferred him to four ships that day, all successfully. I thought the next time services were called, it should be during calm seas.

Wilson was my most experienced crewman, and he said those were the roughest helo transfers he'd ever made in his career.

Tomorrow we anchor near Buckner Bay at Okinawa. I'll be ferrying the admiral to the airbase at Naha.

From the 18th through the 31st of August I've got 15.4 hours of flying time and 59 shipboard landings. My average flight was 15 minutes, most a lot shorter, and I've made two rescues. My confidence is building, and I feel quite comfortable operating a helo in the cruiser environment. But I've also learned not to relax for even a second while flying because disaster may be around the corner.

The secretary of the Army is in Taipei today conferring with Generalissimo Chiang Kai-Shek and his officials. At the end of the visit he announced publicly that it would "ill become the communists to spurn the fair warnings recently issued by President Eisenhower against any attempt to seize the Islands in the Formosa Strait."

Meanwhile the Russians are supporting the Red Chinese. Their newspaper, Pravda, said today that the Soviet Union would give Red China all "necessary moral and material aid" to overcome aggressors, and that American military preparations and aggressive intentions were to blame for the tension in the Formosa area. So this little adventure now involved all the major powers of the world.

Monday, September 1

We arrived at Okinawa, and the ship dropped anchor in Buckner Bay. On the way in, no decision had been made as to whether to fly the helo. So I was decked out in my dress whites to watch the arrival process. Naturally, as soon as I started strolling around topside watching all the activity of entering port, flight quarters was sounded. After a quick change to flight gear, I was off to the airbase at Naha to get mail.

There was none for the *Helena*.

The mail system was all fouled up, but I suspected improvements shortly because the admiral was more than slightly mad at the situation—he wasn't getting any mail, either. Since he controlled all the military activities in the Far East, I figured he'd eventually get results.

On the way back to the ship I found a bonanza—a rain squall. I flew in and out of it several times to give the helo a freshwater washdown, just like a carwash. It washed away a lot of the saltwater and spray that we had been exposed to the past few days.

My first night helicopter landing occurred tonight. The ship sent me back to Naha late in the afternoon with a couple of passengers, including a 225-pound giant commander and a half-dozen suitcases crowded into the back of the helo. We were nearly maxed out on weight. Since the commander was 20 minutes late arriving back at the heliport, I returned to the anchored ship long after dark. The flight deck was lit, but since the ship was anchored in the bay, I felt like it was in a black hole or long dark tunnel with no visual reference on the horizon, and obviously no instruments in the cockpit.

It was a landing I accomplished strictly by focusing all attention on the flight deck and the ship's lighted superstructure. Night landings aboard ships are still not authorized by squadron SOPs (standard operating procedures). But the plans were to weigh anchor at 0100 hours to get out to sea and avoid Typhoon Gracie, and I didn't want to be left behind. When I landed, we latched the helo to the deck with at least a dozen tiedowns.

The XO wouldn't let me eat in his flagship wardroom in flight gear. We had a "discussion" about this, which I naturally lost. The only way I could win was by eating with the stewards. I found it to be a lot of trouble changing from a sweaty flight suit into uniform, then back to flight gear, and sneaking in a shower between changes. My responsibility was to stand by to fly the helo during all daylight hours. I have nights off. But it was a lot of hassle to change for three meals a day and still be ready to go flying. This spit and polish Navy wasn't for me.

Today the Nationalist Chinese reported that their gun batteries on Quemoy shelled a concentration of Communist gunboats and motorized junks. Three gunboats and eight junks were sunk.

Wednesday, September 3

Morale was in the pits for the entire ship's crew. We still had no mail. The admiral was furious. I found out he left a lieutenant commander on the beach in Okinawa with orders not to return to the ship until he brought mail. Probably a lot of homemade cookies shipped by loved ones to members of our crew were lost somewhere in the mail system.

It was a dark night tonight. No moon combined with cloud cover made it pitch dark. The *Lex* was conducting flight ops. All the helos were secured on the hangar deck, and a destroyer was steaming in planeguard formation to the rear of the carrier.

Two F4D Skyray jets attempted to join up in formation near the carrier. The wingman closed too rapidly and in the poor visibility clipped the tail of his leader. Both planes went out of control. The pilots ejected. Both got good chutes.

The destroyer searched in the twisting and choppy seas with giant powerful search lights and miraculously plucked one of the pilots from the dark night seas.

Despite the fact that the other pilot lit flares and fired tracers from his .38 they failed to rescue him.

A destroyer was only about 200 yards away.

A big search was planned for him at the crack of dawn by several carriers and their planes. It'll probably be a waste of time. Chances of surviving in the water all night are real slim—even with survival gear.

It has become clear to me that a person's chance to survive in the water diminishes rapidly after 30 minutes to an hour.

CHAPTER TWELVE

Formosa Strait Crisis

THE CONFLICT IN the Formosa Strait has our ops schedule in turmoil. We were scheduled to go to a place called Saigon in Vietnam, but it was canceled because of the crisis between the Nationalists and the Red Chinese.

From one day to the next, we don't know where we'll be going, how long we'll stay, or what we'll do. As things in the Strait heat up, President Gamal Abdel Nasser of the United Arab Republic (Egypt) said today that Formosa "is and always was a part of China" and now was "under foreign occupation imposed by American direct aggression." But he can't do anything—he's in the middle of a crisis himself with the Israelis.

Meanwhile the British said they're sending a battalion of 600 troops, now stationed in Singapore, to the Hong Kong garrison. But the Brits say they have not monitored any reports that there had been any internal propaganda campaign of Red China preparing the public for news of an invasion of Quemoy or Matsu.

The British voiced their belief that bombardment of Quemoy and the smaller offshore islands was designed to demoralize the islands' inhabitants and garrisons rather than to pave the way for invasions.

The Nationalist Chinese had some small corvettes intercept and sink five Red torpedo boats and set six others afire as the Commies were harassing two landing craft supply ships coming to Quemoy from Formosa. Nine of about 30 newsmen in the Quemoy-bound convoy got ashore by small boats. I felt certain that as the press became more involved and media coverage increased, this was becoming a bigger and bigger event.

Today I ferried Admiral U. S. G. Sharp from the antisubmarine warfare carrier, USS *Princeton* (CV-37), to the *Helena* to meet with Admiral Beakley. I've now operated from every carrier but one stationed in the Pacific fleet. Tomorrow we return to Keelung again, which is becoming our home port.

My roommates are keeping a calendar of the days left on the cruise—158.

Sunday, September 7

The admiral has an urgency about his manner today and not even a hint of a smile. I was scheduled to fly him to Taipei for some conferences with the Chinese, but the weather was so hazy, foggy, and smoggy that he canceled the hop without even consulting me. About an hour later a Marine Lieutenant Colonel pilot on board for the same conference asked me to fly him to Taipei.

"The weather is too lousy to fly," I responded.

"I've flown in worse weather than this," he said, trying to intimidate me.

"I don't care. I'm not flying in this stuff. In a fixed-wing it's easy to fly in bad weather, but in an unstable helo with no instruments for reference, this would be suicide. Even the admiral canceled his hop this morning."

The Marine stormed off and took a car for the 1-hour 15-minute ride to Taipei.

By noon the hot sun had warmed the day and the haze dissipated enough to make flying possible. I made 10 trips between the ship and Taipei transporting virtually all the staff. Something big had to be going on.

President Eisenhower met with Secretary of State Dulles for 2 hours yesterday. The secretary of state said that the president "would not hesitate to order timely and effective military measures" if he decided that a Red Chinese attack on Quemoy and other Chinese offshore islands or the main island of Formosa took place. He also said that "we would not wait until the situation was desperate before we acted to prevent the Communists from overrunning Nationalist forces on Quemoy and Matsu."

Dulles made the following points:

1. Formosa, Quemoy, and Matsu have never been under the authority of the Chinese Communists, and the Nationalists had held them since World War II.
2. The United States had treaty obligations to help defend Formosa, and Congress had authorized the president to ensure defensive related positions such as Quemoy and Matsu.
3. The Red attempt to seize the islands would be a crude violation of the principles on which world order is based, namely, that no country should use armed force to seize new territory.
4. The Reds have been bombarding Quemoy and the other offshore islands harassing their supply lines and broadcasting threats to seize them in Formosa.
5. It is not certain that their purpose is to make an all-out effort to conquer the islands, nor that the islands could not be held by the courageous and purely defensive efforts of the Nationalist Chinese with substantial logistical support from the United States.
6. Because of uncertainty over the Reds' real intentions, the president has not yet made any finding under the joint Congressional resolution on Formosa's defense that employment of U.S. armed forces is required or appropriate to ensure the defense of Formosa. But

the president wouldn't hesitate to do so if it was deemed necessary to protect the offshore islands.

Dulles also said the United States might bomb the Chinese mainland and concentrations of Red forces "if Formosa was attacked or imminently threatened from those airfields" and the same rules applied to the defense of Quemoy.

A reporter asked Dulles whether that was an official warning to the Reds not to attack Quemoy.

Dulles replied, "If I were on the Communist side, I would certainly think very hard before I went ahead in the face of this statement."

The Reds answered by proclaiming all waters within 12 miles of its shoreline as its territorial seas, including Formosa, the Pescadores, and other islands. They also said any ships or aircraft operating in those areas were barred from entering these territories.

The United States recognizes only the 3-mile limit off the Chinese mainland.

Meanwhile the Soviet government newspaper, *Izvestia*, said that the USSR would "in case of necessity aid Red China with all possible means at its disposal just as if its own fate were being decided." It also said the Communist claim to the Nationalist-controlled islands was lawful and just.

This morning the three automobiles and seven boats, including the admiral's barge, that are normally strapped to the aft deck of this giant cruiser, were offloaded and left on the dock.

We obviously are getting ready for battle.

The ship's fuel lines have been over the side all morning, topping off all tanks. Additional food supplies are being loaded. There is a feeling of anticipation in the air.

I've been asked at least 50 times today, "Did you bring us any mail today?" This afternoon I did. A couple of bags had been flown to the Taipei airport, but it wasn't very much.

We leave at midnight headed for the Pescadores Island area—not far from Quemoy and Matsu. We should arrive at daybreak.

It'll be a showdown tomorrow to see if the Red Chinese stop their daily shelling so that the Nationalist Chinese can resupply their beleaguered islands. The Communists previously shelled the islands every other day on even calendar dates, which gave the nationalists the off day to resupply the islands with armaments and food.

Recently the Communists have been shelling the islands every day, and most of the Nationalists' supplies are depleted. The situation for the troops and civilians is becoming desperate.

Word is, if the Communists don't stop shelling, we'll make them do so with all the armed might of the Seventh Fleet plus that of the Nationalist Chinese.

That's what all the conferences have been about the last few days—war planning. The United States is serious about its rhetoric support for an ally. Are the Reds?

A large contingent of newspeople came on board to observe and report what was about to happen.

I met with Dunlap and the crew after chow.

"There's a possibility of some fighting tomorrow. Whatever happens, we'll have to launch and plan on being airborne for as long as we can," I told them.

"The ship won't want the helo around while they're blasting away with their 8-inch guns. The cruiser will be a lot closer to the Chinese mainland than the carriers, so we won't be pulling any planeguard duty.

"We'll probably have three options—fly around the cruiser, stay out of the way and pick up any downed pilots in the area, or go in close and do spotting for the 8-inch gun turrets to help them hit the targets.

"We also become a target then.

"Or we could fly into Red China to pick up any of our downed pilots," I explained.

"We'll just have to wait and see what our orders are."

Two things I wanted were a helo in top shape and a crewman riding along who was very agile. Dunlap was too old, a little heavy, and out of shape. Karanski was a good guy who had only one speed of movement— slow. If we had a problem, I wanted someone who could think and move fast. Wilson was a little older, even though he was only a second-class. He was a thorough mechanic, but also didn't move quickly. That left only Pobst.

"All right, guys. At the crack of dawn I want the helo rigged from a VIP configuration to rescue setup and thoroughly checked for any problems. We'll put in 900 pounds of fuel. I know that's more than we've ever flown with, but we'll need max range. If we can get the ship to give us the right winds for launch, I think we can get airborne without overboosting, but it'll be close.

"Pobst, you'll be crewman. Get prepared and include your survival gear," I ordered.

"See you guys in the morning."

I'd been flying in the bright orange flight suit given me by North American as a result of the rescue of their test pilot, Bill Yoakley. It gave a different look to COMSEVENTHFLT's pilot compared to the rest of the helo pilots.

Now I dug into my gear and took out my ordinary brown flight suit. I figured if something happened and I ended up over the Red Chinese shoreline or walking away from a disabled helo, I wanted as much camouflage as possible. It would have been better to have a camouflage flight suit at a time like this, but they didn't exist.

I unsnapped the brown leather patch, with my wings and name embossed on it in gold, from my flight suit. Then I emptied all the pockets of identification. All my survival school training started to surge into my head. I got out my survival kits, one issued by the Navy and one I put together myself. I put them in the flight suit pockets down by my shins.

Then I got out my .38 (.38-caliber pistol) and bandolier-type holster. I hadn't flown with it yet, but I checked the pistol. I put all tracers in it. In my bandolier shoulder harness and holster I put only about one-third tracers; the rest were copper-coated slugs. I didn't know whether I would need the bullets for signaling or for defense, although I figured if the helo went down along the Red Chinese coast, it'd be pretty tough for one man to stand against the hoards.

I got out my whetstone and honed the blade of my survival knife until it was razor sharp. My roommates watched in silence and pretended to be busy with their own chores. If I was going to go in harm's way, I intended to be prepared.

Then I pulled out a cruise box from under my bunk and rummaged through it looking for all the charts of the area I had taken from the squadron. They were World War II vintage, but better than nothing. I laid all my stuff on my desk, set the alarm for 0530 hours, said my prayers, and drifted off to sleep.

It was a short night. I had breakfast in the wardroom in uniform and went back to my stateroom to change into my flight gear. The gunnery officer, LCDR (Lieutenant Commander) Len Denney, called me to his quarters to confirm what I already suspected.

"Today we're going to escort the resupply ships to Quemoy and Matsu. They're so low on supplies they're just about to starve on those islands. Our government has warned the Chicoms not to fire on the island today. If they do, we'll return the fire," he said.

"If hostilities start, we're going to launch you, and your job will be to fly toward the coast. But stay more than 3 miles off the coastline. CIC will assign you an IFF code so we can keep track of you. If any of our fighters go down in the water along the shoreline or inland, your job will be to go after the pilot and rescue him. You'll be the helo closest to the action."

"Yes, Sir," I answered.

"We'll be busy shelling our targets with the nine 8-inch guns, and we'll be ready for an air attack from the Reds with the twelve 5-inch and fourteen 3-inch guns. I don't know whether we'll launch the Regulus I or not, but we want the flight deck clear so we can if we need to.

"We're going to general quarters at 0900, so you better go see the flag intel officer and see what you can learn about the Chicom defense along the coast, in case you have to fly over the mainland."

"Yes, Sir."

I went to the admiral's staff offices on the Flag Bridge to track down the intelligence officer. I told him the gunnery officer had sent me for a briefing on what I should look out for. I also asked if he had any special instructions or suggestions.

"All the information I have is classified," was his response.

"Yeah, but maybe it's my fanny on the line in a couple of hours. Don't you have any information that would be helpful to me? You know, like what areas to avoid because of heavy ground fire and stuff like that?" I challenged.

"I told you, it's all classified."

I was furious, but after 5 minutes it was apparent that any further conversation was a waste of time. I went down to my stateroom, collected my gear, and headed for the fantail and crew. They had the blades spread and booted and had the helo filled. The seat was rigged forward. They were ready to go. They had even procured box lunches for us.

We sat around and talked as general quarters sounded. The ship's crew raced around the ship. All hatches were closed and dogged shut.

The few crewmen outside the protective steel superstructure, manning the antiaircraft guns and taking care of other responsibilities, all wore kapok life jackets and gray steel helmets.

The turrets were swiveling around as the crews simulated tracking and firing at targets.

Flight quarters had already been sounded. I was strapped in with all my survival gear, gun, knife, and bullets and even wore my helmet. Pobst was ready in the seat behind me. He had no pistol or special survival kits. The squadron didn't deem it necessary to give white hats the same survival equipment it did to pilots. Too bad I hadn't accepted the pistols from my AD rescuees.

The other three crew members stood by my open window ready to monitor the fire extinguishers, start the engines, pull the tiedowns, and launch me. There was occasional small talk. My mind wondered about the lack of any intelligence briefing.

I finally came to the conclusion the intelligence officer didn't know any more than I did, but he didn't want to admit his ignorance, so he used his flimsy "classified" excuse.

If someone went down, all I could do is be vectored to the downed pilot by his wingman for the pickup. I just hoped I was in the air long enough to burn off enough fuel so I could carry the added weight of the rescuee. There was no quick way to dump fuel from the HUP tanks except fly it off or siphon it out.

The ship continuously maneuvered and zigzagged to prevent becoming an easy target. It was cruising at high speed. The shafts were really turning and churning. I could feel the vibrations; they seemed exaggerated on the fantail.

Occasionally a reporter or correspondent would wander back to see what was going on at the stern of the ship. The guy from Associated Press said we were escorting some Nationalist Chinese supply ships and daring the Reds to fire at us. He wondered how close we could come to the Nationalist island before running aground. Running the Red blockade, we escorted the ships almost to the muzzles of the Communist artillery.

According to international law, we had every right to be where we were since we were more than 3 miles off the Chinese coastline.

I looked up at the skies and saw patrols of jets, some fairly low and some higher. I wondered who belonged to the white scratches or contrails overhead. I couldn't see the big picture, but it was easy to tell there was a lot of military activity going on, and the admiral wasn't off hiding somewhere. He had his flagship square in the middle of everything.

I listened to UHF radio frequencies, but didn't learn much more of what was happening. There were no emergencies on guard channel.

The day wore on. Then the tension relaxed a little. There was no action. Evidently the Communists had backed down.

The entire ship stayed at general quarters until 1500 hours. Then orders were given to back off to condition One Alpha (1A), which kept everyone at battle stations, but allowed the opening of some of the hatches for better ventilation below decks.

The AP (Associated Press) guy wandered back again and said the Reds blinked.

They didn't fire and allowed the Nationalist resupply ships to get to the islands.

This was as close to the brink of a war the United States had been since the Korean War ended 5 years earlier.

Since the helo was useless for normal hops with so much fuel, the crew siphoned out the excess fuel and brought the tank from 900 to 450 pounds.

The ship turned and was headed east. As we approached the Pescadores Islands in the Formosa Strait I made two trips to a local airfield to transfer war correspondents. Personnel included men from Associated Press, *Time*, *Life*, and *The Chicago Daily News*. From there they'd catch a Navy fixed-wing aircraft and fly back to Taipei to file their stories for the world to read about the day's events. Looking down, I could see that every crossroad on the island has a pillbox and armed guard. Troops are billeted in private homes and temples. They carry all their possessions with them. The main style of combat footwear is American "sneakers."

Monday, September 8

Yesterday, with the threat of military force, the United States had forced the Red Chinese to stop their bombardment of Quemoy and Matsu. The Communists had lost face. In an effort to reestablish some credibility, they again started the artillery bombardment and sank some Nationalist Chinese supply vessels approaching Quemoy inside the 3-mile limit. There was no effort to fire on the American warships in the area.

Aboard ship operations are pretty much back to normal. We're cruising back to Keelung to pick up the barges and cars left behind Saturday.

We stopped along the way and anchored near the southern tip of Formosa to watch a combined landing exercise with U.S. Marines and Chinese

Nationalist troops. We guessed they were on alert yesterday and had nothing to do, so this joint exercise was popped on them.

The afternoon seas were real rough. The frothy blue-green waves undulated along the side of the ship as we steamed along. The strong wind whipped a mist from the whitecaps on the superstructure from the starboard side of the ship. The ship wanted me to fly some newspeople to the carrier. I refused because of the treacherous weather, so I guess the newspeople will stay on the ship for another day or so.

The next item on the agenda was replenishing food and supplies from a supply ship. As the two ships steamed along in formation, the water between the vessels frothed up. Part of the time the supplies were dunked under the violent whitecaps. After a short while there were orders to stop. Half the cargo of pallets and boxes being highlined between the ships were lost during the transfer because of the rough weather.

Refueling was the next thing to try. Things got even worse.

At 15 knots the first pass to come alongside the tanker USS *Ponchatoula* (AO-A48) on a parallel course failed. The objective was for the ships to steam side by side at 15 knots, separated by only 75 or 80 feet of wild, boiling ocean. A network of lines and oil hoses was to link the two ships.

Three thousand gallons a minute can be pumped through each of the heavy transfer hoses of the tanker. Our ship was going too fast, so we went around for a second pass.

Just about the time we were alongside, the tanker's steering system failed, and they turned into us. It was a minor collision, but everyone on the ship from the engine room to the signal bridge felt the jolt. The bang must have really frightened those inside the hull where the ships hit. No injuries except for a big dent in the side of our ship. Repairs will have to be made, and estimates are it'll take 2 weeks. All this occurred with the group of newspeople still on board to observe the captain's embarrassment.

At least I felt vindicated about refusing to fly in the unsafe conditions and didn't get my helo banged up. But the salt spray corroded the brakes on my helo. The corrosion caused a hydraulic leak and made a mess on the wooden flight deck. There's no flying at sea without brakes. It would be impossible to stop the helo from rolling forward or backward with the ship's movement.

Tuesday, September 9

The replacement brake assembly from the ship's supply system was in worse shape than the one already on the helo. My crew stripped parts from both brake systems to finally get one that would work.

Today I had my roughest flight ever in a helicopter. We anchored in Keelung. The admiral and his flag lieutenant, LCDR F. Brown, my old friend from Miramar, wanted to go to Taipei. True to form, he chose to fly only in bad weather conditions. The visibility and rain were okay, but the turbulence was tremendous. It took every ounce of skill I had developed to keep control as the helo bounced up and down, jerked and jolted sideways nearly tumbling out of the sky.

At 700 feet, going between some mountain passes I wondered how the flapping rotor blades held together. I was so busy trying to control my machine, I didn't have much time to be afraid. I even considered turning off the control boost to make the helo more stable, but decided against it.

There was no intercom talking since my passengers didn't wear headsets. I was moving the cyclic in circles trying to keep us level, and the up/down drafts had me pushing and pulling the collective like a water pump to maintain stable altitude. It was a continuous fight.

The admiral was sitting in the frontseat next to me, and LCDR Brown was sitting behind me. About 5 minutes into the turbulence, roughly half way to Taipei, one of those two pilots became sick, scared, or unsettled. A foul smell enveloped the inside of the helo. It was suffocating, and I started to choke on it. I couldn't tell who, but one of them had probably soiled his dress white pants. In spite of the tension I was under at the moment, I had to laugh to myself.

We finally got to the Taipei Airport. The weather had smoothed out a little bit. I looked at the seat of the admiral's and his aide's pants as they deplaned, but I couldn't tell which one was guilty. I didn't see any evidence on the outside seat of either of their white pants as they crawled out.

The admiral came over to the window.

"Lieutenant, you don't need to fly back if you feel the weather is too bad."

"I'm sorry for the rough ride, Sir."

"I'll drive back tonight so you won't need to come back for me."

Wednesday, September 10

We joined up with the *Midway* and *Lexington* today. This was our first chance to see the crews from the *Midway*. Three planes crashed in a period of about an hour.

It was a disastrous day for the carriers. It started off when an F8U from the *Midway* caught fire in flight. There wasn't anything the pilot could do except eject. He got out safely and, in spite of a tangled chute, made a successful water landing and was in his one-man raft when Lieutenant jg Dick Browning picked him up with his HUP in an uneventful rescue.

About 15 minutes later an AD from the *Lexington* lost his propeller when the prop governor failed and the engine oversped, throwing the propeller off and spraying oil all over the windshield. The pilot thought about ditching, but couldn't see, so he bailed out. He was badly banged up in the bailout and severely injured.

The *Lex* helo piloted by Ensign Hale was busy transferring the chaplin to a destroyer at the time of the accident. He raced back to his carrier to dump off the chaplain and pick up a crewman to make the rescue. Meanwhile the HUP from the *Midway* piloted by Lieutenant jg Sam Anderson raced for the rescue, beating out Ensign Hale. There's a lot of pride swallowing that takes place when a helo crew gets beat out for the pickup of one of the pilots from his own ship and a lot of swaggering by the other ship's helo crew. In this case it was doubly painful, because the squadron of the rescued pilot sent over an appreciation cake to Lieutenant jg Anderson—via the *Lexington* helo.

Before an hour was over, tragedy struck again. The seas were rough and the bad weather caused an erratic pitching deck. An F3H Demon piloted by the commanding officer of the squadron on the *Midway* was attempting a landing when the aft end of the ship pitched up higher than expected, causing a ramp strike.

The plane slammed on the deck, shearing off the landing gear.

It slid out of control up the angled deck, flames spewing out the exhaust. It then slid over the side of the ship into the water, creating a giant splash.

The plane sank immediately.

The pilot never got out. None of the flight deck personnel were injured.

Thursday, September 11

This VIP business isn't all it's cracked up to be. I logged in 19 shipboard landings and 3.2 flight hours. Every time I go to a carrier, there is never a chance to fly planeguard. I just sit around, sometimes for hours waiting to haul people around, while all the other helos are out flying.

Of course, my helo is outfitted differently than the planeguard helos. They're dirty, greasy machines. Mine is like a Cadillac, all polished up with white seatcovers and special white sleeves for the lapbelts. We even have white covers that fit over the shoulder harness so no one ever gets dirty flying in the *Helena* helo.

When the admiral flies, we have a plate with three stars that we slip into a holder on the outside of the aircraft so that everyone can know he's on board.

Dr. Cabrera called me to sick bay today.

"Mr. McKinnon, one of your men has the clap. I want you to be aware of his medical condition. He'll need treatment."

"What's the clap," I asked naively.

"VD, gonorrhea," he replied.

"Where did he get it?" I asked.

"In Keelung, he thinks."

"Can it be cured?"

"I'll take about 3 weeks."

I restricted my infected sailor for 3 weeks, but what I didn't tell him was that with our current fast-paced tempo of ops we probably weren't going to get much liberty in port the next 3 weeks, anyway.

Morale slipped to a new low on the *Helena*. The policies of both the CO and XO haven't gained them any love from the ship's company. Their orders seem arbitrary and without any logic or reason. I have my usual daily arguments about flight ops, but I've got them over a barrel. I don't have to fly my machine when in my sole judgment I think conditions are too dangerous. It really frosts them.

There are no written rules about pilot judgment in determining what are satisfactory conditions for helicopter operations. There's one paragraph in a COMNAVAIRPAC instruction book that leaves the final decision to fly in the hands of the pilot. There ought to be some written operational rules to protect helo pilots from harassment of ship's company (blackshoes) who don't understand aviation operations and what is safe and what isn't.

Things at Quemoy and Matsu have quieted down. There's still some shelling going on, but so is the resupply. The tension from last Sunday's event has abated.

Admiral Beakley says rough seas will prevent a Chinese Communist invasion of Formosa after September 15.

"I would say if we get through September 15, Formosa is not too threatened," the Seventh Fleet commander said at a news conference. "The water gets too rough for invasion in the Strait after that.

"August is the best month for anything to happen," he said. "In September the weather starts getting bad and October is worse. The Chinese Communists do not have heavy amphibious forces such as we have. When the water gets rough there is no danger of a Strait crossing. After September, they can't try it again until about April 15, and then only through the end of May when the window of opportunity closes again," he told a group of newsmen on the ship.

Friday, September 12

Assistant Secretary of the Navy Armstrong was my passenger today. He was wearing some strange green flight suit (it must have been Air Force) and a bright red baseball cap. He is a baldheaded guy wearing wire rim glasses who seems to be on a joyride out here.

As we lifted off the flight deck of the *Helena* and flew alongside the ship, he remarked, "Sure a lot of grass growing on the side of the ship. When do they clean that off?"

I thought to myself, we've been at sea 3 months with all kinds of wild things happening, very little mail, constant convoys and nearly going to war, and he's concerned about grass growing on the waterline of the ship.

"I guess it grows fast out here at sea," I diplomatically responded over the intercom.

Ship operations were working at cross purposes. We have to fold our rotor blades and spread them at least three times a day—that's more than a carrier helo does in a week. They're wearing my crew out for unnecessary reasons and poor planning. We should have to fold and spread only once a day. The ship feels having the helo on the flight deck blocks the area's usage for more important things. The rotor heads are getting more wear and tear from the folding and spreading than from flying.

Everything has to be done in emergency speed and then we sit around for an hour until launch time. I'm frustrated by it and tried to get it across to the XO.

The *Midway* lost another pilot last night. They were doing night operations, and a young lieutenant made pass after pass trying to get aboard in his F3H-2N Demon. After the seventh unsuccessful attempt to catch a wire, he was about to run out of fuel. He was instructed to climb up in front of the ship and eject while he still had control of his aircraft. He did so. Since helos can't fly at night, the destroyers crisscrossed the area where radar plots showed the plane impacted in the sea. The carrier slowed and came to a stop with hundreds of men manning the rail looking for the pilot.

They never found him.

Night landings are the pits. At this rate they're going to run out of airplanes and pilots soon. Jets are dangerous, and I'm thankful I'm flying a helo at this stage.

Saturday, September 13

We're in Naha, Okinawa today. The island is long and narrow with sandy beaches and surrounded by coral reefs. Some of it is flat, and other parts have high cliffs.

The ship had a firing exercise planned for a couple of days. To protect the helo from the concussion of the gun blasts, I've lined up some hangar space on Naha for us to work on the helo.

Weather sure can affect a lot of lives.

All liberty was canceled. Typhoon Helen is moving in, and the captain feels the best protection for a ship is on the high seas. Anchored in port where the anchor could drag or anchor chains break, his ship could be beached.

I flew back in high winds, but in the harbor the deck wasn't rolling too much. I sweated the shutdown of the blades without busting one against the side of the helo. As we put out to sea, the wind and sea swells really got us rolling.

While I sat in my room trying to adjust to the ship's roll the captain's orderly came with a message, "The captain wants the admiral investigating the collision with the tanker flown back to the beach."

"I can't fly in this type of weather," I responded. "It's just too rough and it's getting so late, I probably couldn't get back before dark anyway."

"You better tell that to the captain yourself, Sir," the Marine guard replied.

So I was off to the bridge.

The captain was insistent. I had him look at his own instruments, the anemometer and the inclinometer. He could see we were rolling like crazy by looking at the horizon. He'd canceled everything himself and was leaving port because of the typhoon. Now why the pressure on me to fly in this lousy weather?

He wanted to get rid of the inspecting admiral. Since I wouldn't fly him off, now he was stuck with him. The admiral was really pissed off.

So was I. It always comes to a fight over marginal operating conditions. I'm sure if I had an accident out here, the ship's company wouldn't come to my rescue or take any of the blame. The fault would all be on my shoulders.

Even the admiral's staff is angry that we can't get rid of the investigating admiral. They probably don't want him snooping around anymore than he has to. It'll be a few days before the bad weather will allow for his transfer from the *Helena*.

President Eisenhower expressed belief today that the Red Chinese were using the Formosa Strait offshore islands as a place to test the free world's courage to risk the aggression.

"If the Red Chinese have now decided to risk a war over Quemoy, it can only be because they and their Soviet allies have decided to find out whether a threatening war is a policy from which they can make big gains," the president said today in a radio and TV address to the nation. He also declared the Formosa Strait situation serious, but not desperate or hopeless and asserted, "There's not going to be any war."

Of course we're sitting out here ready for combat at a moment's notice. It's going to be interesting how this continues to develop.

Meanwhile Khrushchev has taunted that the United States and Nationalist China were to blame for the renewal of tension in the Formosa area.

Eisenhower warned Khrushchev that American forces are fulfilling treaty obligations to Nationalist China to assist in the defense of Formosa and the Pescadores and that no upside presentation of the facts by him can change U.S. policy.

Former President Harry Truman has been supporting President Eisenhower in his tough stand.

Obviously our continued exercises are to prepare us for combat.

I think we're headed for Yokosuka after we dodge the heavy part of the typhoon. But information is classified, even to us on the ship. And we don't even have anyone on the outside world to talk with.

Sunday, September 14

I was wrong. We moved to Formosa, and I flew the admiral to Taipei. It was a rare flight in clear, calm, smooth skies. It was such a peaceful flight that I could have stayed up for hours savoring the beauty of it all.

Looking down I could see the lush green terraces of vegetation, red clay tiles on the roofs of light brown homes with wisps of bluish smoke rising from rice fires, railroad tracks snaking through the valleys, and pottery kilns scattered along the hillsides.

We remained in port for just 6 hours so that the admiral could go to meetings. Plans aboard the flagship are always in flux. The poor captain of the *Helena* is nothing more than a boat taxi driver.

We have unexpectedly left behind all the admirals, captains, and officer barges as well as the automobiles we normally carry aboard the ship. This, of course, indicates we'll be firing the ship's guns soon. Whether it will be for practice or more convoy duty, no one knows.

We have navigated out of the path of Typhoon Helen. Some aviators on the admiral's staff have offered to put me on the SNB Beechcraft flight schedule when we are in Japan. They use the aircraft out of Atsugi to get in their proficiency flight time. This should give me a little instrument time plus keep me current in fixed-wing aircraft.

The ship's XO and ops officer must be gaining respect for my flying. They now call me their CAG (commander of the air group). That's a title reserved for the man in charge of all the planes on a carrier. I guess I am the CAG on the cruiser—I'm in charge of all our aircraft—the one helo!

Chiang Kai-Shek made a statement in today's New York *Herald Tribune* in a story by Joseph Alsop, whom I taxied around out here. Alsop was a middle-aged tall balding guy wearing a tan vest with pockets everywhere to carry film and camera equipment. Chiang said the Quemoy garrison and population cannot be sustained by convoys landing under the withering Communist fire and that the "one real solution to the problem" was to "attack it at the source" by knocking out the Reds' mainland batteries.

"This kind of retaliation had silenced Red attacks in 1954," he said. Of course it would involve deep involvement from the U.S. forces. It seems this is his attempt to get the United States involved even more. He's a real hawk.

There was some exciting action yesterday. Quemoy and Matsu had been isolated because of the heavy storms and typhoons. An LST had zigzagged through the artillery fire from the mainland batteries. The shipment of supplies was augmented by an airdrop of supplies from four Nationalist C-46 planes, so obviously blockade running techniques were being developed. But they need to get through enough supplies for 100,000 men and a civilian population of 50,000 so they won't starve.

The Red Chinese also said that about 5 days ago U.S. Air Force U-2 spy planes had flown over their territory to do reconnaissance and they were upset about it.

Wednesday, September 17

We had 8-inch gun firing practice today about 100 miles from Okinawa. They fired at a 500-yard-long rock. It was a wild scene from the helo. I was ordered to take with me Captain Cliff Johnson, the O-in-C of the Marine detachment aboard the *Helena*. He's a typical Marine infantryman with his blond hair cut so short he appears to be bald.

His job was to spot and call back to the ship where the shells landed. The procedure was for the ship to radio "fire" when they sent the 8-inch shells heading for the island from 10 miles or so away.

We would fly at about 1000 feet abeam the island looking for dust or splashes in the water. Then Captain Johnson was supposed to call back corrections to allow the ship to bracket in and hit the island.

To begin this ordeal he'd never been in a helo. It was a hell of a project to brief him about his survival gear, to strap him in, and get him checked out on the radio. I put him in the VIP seat next to me. Marines think in sequential fashion like 1,2,3,4,5. They have a hard time going from 1 to 4 and skipping parts 2 and 3.

We finally took off. At this point I found out he hadn't been briefed by the ship's gunnery people on how they were going to coordinate the day's activities. We orbited around for $^1/_2$ hour or so killing time.

Then without warning, the ship called over the radio "fire." Naturally we were in a turn away from the large rock, which protrudes about 500 feet above the Pacific Ocean and is called an "island."

I don't know how long it takes shells to travel 10 miles—the distance the ship is from the rock. I do know it's impossible to hover at 1000 feet while we wait. We had to circle or fly a straight line and then hopefully be in position to watch the shells hit.

My Marine friend is going bananas trying to loosen his seatbelt so he can look out the window. He's squirming all over the seat to get a better look at the rock. All the while I'm trying to maneuver the helo. I wonder to myself how long it is going to take before this guy gets sick from all the motion.

I maneuver the helo around in time and we see splashes in the water about 1000 yards short of the rock.

"Short, short—two clicks up," Johnson says as he flips on the mike switch and radios back to the ship.

Two clicks? I thought. What does that tell the ship? They aren't shooting M1s. There must be a better system.

The ship radios back with a puzzled, "Roger." I think they didn't want to admit they didn't understand the message coming back from their spotter helo.

"Fire," came over the radio again.

We spotted the shells falling 600 or so yards short.

"Add two more clicks," radioed a confused Johnson as he wiggled around. The ship still didn't know how far they were missing the target, but whatever increment they were using to close the distance to the rock, they were getting closer to make a direct hit.

"Fire," cracked the radio.

The splashes were now past the rock. Firing 8-inch guns at long range while bobbing on the seas and trying to correct for the forward speed and the roll of the ship sure was difficult.

I felt the best technique would have been to give the estimate in yards of where the splashes were in relation to the rock instead of "clicks," and let the ship figure out how to make the correction.

We spent about 45 minutes going through this spotting procedure. The ship did eventually hit the island, which registered with clouds of guano deposited by seagulls and birds.

After the session was over, I dropped down, and we examined the foxhole-size craters the artillery created on the island. It was obvious that those shells would carry an awesome impact when they hit the enemy.

On landing aboard the ship, Captain Johnson was summoned to the bridge for a debrief. I heard it wasn't too friendly. Word came back to me that the helo pilot would do all the spotting in the future.

Friday, September 19

Having a helo aboard proved its worth again.

Yesterday the ship had an antiaircraft gun practice. Because of the concussion factor and fireline restrictions, it was impossible to have the helo on the fantail. I was launched to a carrier and forced to shut down and wait until the exercise was over.

Today I got to orbit around and watch the small target drone they were trying to shoot down. It eventually ran out of gas, popped a parachute, and drifted into the sea. I spotted it and marked its location with a smoke bomb about 5 miles from the ship. Yesterday they lost the drone. We found it today with the helo—so we earned our keep.

I decided to start keeping a notebook of the lessons I've learned and things I ought to do as a boss if I ever find myself in a situation with more than four crewmen. Those thoughts might provide some helpful insight to understand what those under me think.

We completed a 60-hour mechanical check of the helo last night. All maintenance work is done at night with flashlights. It's tough on the crew, but I want us always ready to fly during daylight hours. It's hard enough getting flight time, without having the aircraft downed with maintenance problems during the day.

The engineers are having problems with the boilers. They can't be fixed while we're under way, so it looks like we'll return to Keelung first thing tomorrow.

The boiler problem has been continuous. We've been on water hours for more than 10 days. This means the evaporators can convert saltwater to freshwater for each man to have only 20 gallons of freshwater per day for cooking, laundry, drinking, and showers. We're limited to Navy showers, which amounts to turning the shower faucet on and off, then soaping down, and finally a quick rinse. The master at arms has a guard stationed at the showers to be sure orders are followed—officers included. The drinking fountains are turned on for an hour, every 3 hours, to let the crew have a drink.

This also affects our ability to wash the saltwater off the helo. We're now given buckets of water instead of hoses. We are allowed to use the hose every third day.

I don't know if we'll ever get back to Yokosuka. Keelung seems to be our home port. I go ashore rarely. It's a hellhole made up of bars, prostitutes, and sailor ripoff joints. We've never been in port long enough to leave the city or to visit other parts of Taiwan or even Taipei to see how the Chinese really live.

While we were sitting around the helo waiting to deliver some mail today, Chaplain Keeney and I got into a 45-minute discussion about morals, manners, and principles. We agonized over the behavior of some of the young sailors and even the officers when they hit port. My conclusion was a lack of parental training and certainly a lack of self-discipline. He felt they didn't adhere to their faith.

The effort and significance to provide some type of faith for our military personnel in today's world has been diluted by the actions the U.S. Army has taken by redesigning its chaplain's crest. The cross has been removed as the symbol of Christianity. Gone, also, are the tablets and star of David that represented the Jewish faith. In their place are a depiction of the sun with its rays, which refers to the presence of God in nature, and other symbolic drawings that skirt our Judeo-Christian heritage. And no one seems to object to these changes.

At least with the flagship we have hit port a few times. The guys on the *Lex* are really beefin'. They've been at sea 44 straight days. Last time they were in port was for 10 days ending July 16.

All the activity with the Reds and potential for war has attracted the big names in today's *Who's Who of Journalists*. Today's visitor was Stewart Alsop, whom I brought to the ship for an interview with the admiral. It was his brother that was here less than a week ago. Both are columnists.

Tonight we're refueling again from the tanker. There was no collision. Today I got in 4.0 hours of flying, including 15 minutes planeguarding on the *Midway*. The *Midway* helos have been having more than their share of mechanical problems. On the average, a HUP requires 15 hours of maintenance for every flight hour.

The Red Chinese have a new tactic to grab headlines. They are staging protests. They claim 302 million people in Red China protested the United States' involvement with Formosa. They called it "the biggest protest campaign in world history."

The war of words continues over the Formosa Strait. Soviet Premier Khrushchev sent a note to President Eisenhower saying "the U.S. forces must leave the Formosa area or suffer expulsion by Red China."

Eisenhower sent it back and told Khrushchev the note was unacceptable under international practices and concluded it was full of false accusations and abusive and intemperate language. He wasn't going to be threatened by some political note from Khrushchev.

Khrushchev's note demanded that American naval fleets be withdrawn from the Formosa Strait, that American soldiers leave Formosa and go home. If the United States didn't do that, then no other way was left open to the people of China except expulsion of armed forces hostile to it from its own territory.

Eisenhower's response also included the fact that the United States was demanding an immediate cease-fire on Quemoy and Matsu.

Saturday, September 20

Today was a busy day. I flew 6 hours and made 21 shipboard landings. I delivered mail and cargo from the carrier to all the ships in the task group. This included landing on a cruiser and hovering over the fantails of destroyers while my load was delivered by cable.

I wonder if the monotony of long at-sea periods forces people to make mistakes. We made one by delivering some ship parts and supplies to the wrong destroyer. Delivery to a destroyer is done by stacking a bunch of boxes and sacks of mail in the back of the helo. The crewman has to read the name of the ship on the label and put it in a bag, while I read him the name painted on the stern of the ship over the intercom. Then he lowers the cargo by hoist to the ship.

My crew seems not to be paying attention. They put some wrong packages in the pouch to the wrong ship.

When you love to fly, all the challenges of these shipboard ops keep you excited. There's a certain mental competition that takes place in your mind about trying to be the best. But the helo vibrations and need to be constantly alert because of the precision required for shipboard landings on a rolling and pitching deck caused fatigue to set in today. It got worse as the day wore on because all I had to eat was a white bread baloney sandwich and a Coca-Cola. Not much nourishment.

The basic way to navigate from the carrier is with a laptop maneuvering board that plots the ship course, speed and time, and your aircraft course, plus some corrections for winds. It's a big board that sits on a pilot's lap.

An AD pilot from the *Midway* somehow didn't keep track of his maneuvering tonight, got lost in the dark, and ditched his aircraft before he ran out of fuel. He was extremely lucky to have made a successful night water ditching and then get out of the aircraft without injury.

Tomorrow morning, I'm scheduled to take the admiral to the carrier. From there he'll fly to Taipei to have a conference and fly back in the afternoon. I already took him to the carrier this morning just after sunrise. Whatever our efforts are to oppose the Chicoms is taking a lot of coordination.

I'm feeling very comfortable with cruiser operations. I have 34 hours of flying this month. That's the most I've had in any month since completing flight training in Pensacola 9 months ago.

Sunday, September 21

The war continues between the two Chinas. The Nationalist Chinese government claimed that its F-86s shot down five Red MiG-17s and sank three Red torpedo boats 4 days ago.

Supposedly the Nationalist Air Force is staying out of combat over China to deny the Reds propaganda advantage in case one of their pilots got shot down over the mainland and became a POW.

The Nationalists also estimated that the dead and injured from the Red bombardment of Quemoy from 23 August through today totaled 3000 civilians and 1000 military personnel.

Meanwhile my admiral's boss, Admiral Harry D. Felt, commander in chief of the Pacific, was in Taipei yesterday and announced that our Seventh Fleet and the airforces at Formosa were "very, very, strong" and "quite adequate" to deal with any drastic action the Communists might take. He met with Chiang Kai-Shek.

Monday, September 22

We arrived in Buckner Bay in Okinawa. I flew more than 6 hours ferrying the admiral and his staff. All of a sudden the ship needed the admiral's signature on some documents, so I picked up an aide on the anchored ship and delivered him and the papers to the admiral. I landed on a baseball diamond on the Marine base on the island.

We learned today that Vice Admiral Frederick Kivette will relieve Admiral Wallace Beakley as COMSEVENTHFLT September 30th in a formal ceremony aboard the *Midway*. Beakley has been a grouch. Everyone on the ship is glad of his leaving. We don't know whether his leaving the Western Pacific, or the constant pressure of the Taiwan situation, is wearing him down.

Tuesday, September 23

It was an extremely rough day at sea today. We're on the fringes of Typhoon Ida, causing lots of rain and rough choppy seas. All flight ops on the carrier were suspended. Landing the jets is dangerous because the rough seas cause large pitches in the carrier fore/aft movement, which causes the planes to slam down on the ship extremely hard. Occasionally this pushes the oleos up so hard that they puncture the aircraft wing.

An ammunition ship accompanying our task group had an emergency leave case. An enlisted man had some type of emergency concerning his wife back in the States. They wanted to get him positioned on a carrier so that he could fly on the COD to some shore base and then fly home.

The carriers wouldn't let their helos fly because of the rough seas. If it's too rough on a carrier, it's easy to imagine what the high winds and rolling swells are doing to our smaller and less stable cruiser. I was asked to fly over, hoist the sailor aboard the helo, and deliver him to the carrier. We're developing the reputation of flying when no one else will.

"We'll try it," I told the gunnery officer.

"But if you guys can't give me a course that neutralizes some of the wind and roll, we'll have to scrub the effort."

"You call the shots, Dan," he replied.

They finally settled the ship on a course that I thought would be okay to engage the rotors. As I engaged the friction switch, the ship lurched and the speeding rotors flapped around severely, but somehow missed slapping into the side of the helo. For a moment I worried about the rotor disengagement and the shutdown perils I faced at the end of the mission. But the rotors were gaining speed. My commitment was too strong to stop now.

The events of the moment demanded my full attention. With the rotors up to speed and the jaw switch engaged, there's not much of a problem of their hitting the helo pylon or fuselage, but the gyroscopic effect of the rolling ship requires the helo to be tightly tied down.

To get launched in these conditions requires close timing. As we started to roll to a level position, I signaled for the tiedowns to be pulled quickly and lifted off with a lot of up collective. I raised the helo higher than normal to avoid a tail strike against the ship's deck. I was nervous enough trying something no one else would do without margins to be safe.

It was a bumpy ride for both me and my crewman, but we made it to the ammo ship. The stern was bobbing up and down with exaggerated movements. Fortunately the ship had the wind at 45 degrees across the bow, which kept the stack gases from causing additional turbulence for the helicopter.

I gradually nosed in a little high with the hatch open. My crewman was ready to drop the hoist cable and sling down to the sailor.

I didn't know what the emergency was at home, but felt if his wife or whoever wanted him home saw what we were about to do, they might have decided they could handle their problems alone.

Salt spray was whipping around everywhere from the wind and rotor blade down force. I came to a hover behind the ship and gradually eased forward. Looking down, I saw that the ship's crew assigned to this task were tethered to the rails and were wearing dungarees and kapok lifevests and were drenched from the seawater spraying over the ship's rails. The poor sailor to be transferred stood out in his dress blues clutching a small bag.

The fantail kept rising and falling. I hovered high—at least 25 feet above the deck.

I could get only the nose of the helo over the ship. The back half remained over the water.

I had to be careful.

My rotors were about 5 feet from some halyards, cables, and radio antennas on the ammo ship. If I got careless, they would have clipped the tips of my forward rotor blades off, and I could have easily ended up a fireball on the fantail. Or if the ship was lucky, I'd have gone into the water, nose first with my crewman, and probably pulled in the unlucky sailor with us.

It took a lot of control and maneuvering, but we finally got the sailor into the helo, backed off, and headed for the carrier.

The captain of the ammo ship was impressed with our operation and sent me a "well done under adverse conditions" over the radio. The best part was that it was on the *Helena*'s frequency and they heard the message, too.

The landing on the carrier was uneventful, although I didn't shut down the rotors. I razzed the carrier helo crew about me doing their work for them as they gathered around to say "Hi."

"What's wrong with you guys? Too chicken to fly today?"

They were envious and laid all the blame on their ship's air boss. Fortunately, the *Helena* was in good winds, and my recovery and shutdown went smoothly.

Wednesday, September 24

The battle over the Formosa Strait continues. The Nationalist Chinese government claim their F-86 Sabre jets shot down at least 10 MiG-17s today without a single loss.

U.S. Air Force advisers on Formosa upheld the Nationalists' claims using as evidence gun camera pictures and other communications.

Evidently the Nationalist Chinese are getting supplies to Quemoy on private junks and military ships.

We experienced more gun firing at some islands today. The spotting task went much smoother than before, and the guns weren't too far off the target. The ship was 13.5 miles away and the misses were only a couple of hundred yards off target. The seas continue to be rough with strong and gusty winds.

I flew 2.3 hours and have 41 hours total this month. I may reach 50 by the end of the month, which would be a record. A normal amount is 15 to 20 hours, sometimes not even that much.

My relations with the bridge have not improved. There are still arguments about flight conditions, in spite of the great cooperation for yesterday's launch to the ammo ship. They still call me to the bridge and chew me out and shout, but I usually argue and don't let them get to me if I feel my operation was correct. By now we should have developed a better routine and understanding.

The fellows standing officer of the deck watches pretty well accept my decisions and explanations. My biggest problems are with the operations officer and XO. The captain doesn't say a word. He doesn't need to; he had others do it for him.

Now we've entered a new phase in our relationship with the bridge. They are tired of our relaxed response to their "urgent" needs. My determination is not to sacrifice safety for expediency, particularly when there is no emergency and my life is on the line.

I've written a section in my cruise report to the squadron, urging them to use higher authority to issue standardized rules under which helicopters can operate. It isn't fair to pit a Lieutenant jg against a captain in an argument over safety. The Lieutenant jg bears the ultimate responsibility for his actions, and the captain does not get his way because of the disagreeable junior officer.

I'm the only helo pilot out here that will fly under absolutely terrible conditions. Part of the reason is that I have developed greater finesse and control with the helo operating within the narrow confines of *Helena*'s flight deck. Another part is because of my passion for flying and confidence in what we're doing.

The newest strategy is to call my direct boss on the ship, the gunnery officer, to the bridge, chew him out for my actions, and expect him to get results from me. It makes the guys on the bridge feel better, but the results are the same. I'm still the guy who is responsible for the helo. I've got to do it the way I feel safe.

The CO and XO of this ship have their own ideas on how to handle men. They insist on building a fire under everyone. It seems to me a pat on the back and a friendly suggestion are enough, and to get tough if necessary. Everyone knows what his responsibilities are. Give the guy a chance before chewing him out.

I'm still barred from wearing flight gear in the wardroom, but it's worked out fine. The stewards appreciate their mailman, so I eat with them in the galley area a lot. They always serve me the best food and cook special steaks, and I get all I can eat. I always praise the delicious food, and they bend over backward to get me food any time of the day or night. It's not a bad deal, and they are good guys to have a good relationship with. It's the little things that can make life worthwhile aboard ship.

This Saturday, Sunday, and Monday we're scheduled to be in Buckner Bay. We have a lot of repair work to do on our bird, including replacing both brake assemblies that have corroded. Saltwater is really rough on aluminum parts.

Thursday, September 25

Today was a good day. We flew the admiral to the carrier, where he took a twin engine A3D jet bomber to the Philippines about noon. I stayed aboard the carrier and let one of the other fellows fly planeguard. During this time I bought a new helmet and some other flight gear from the carrier's supply system.

It was fun sitting around in nice air-conditioned, roomy, two-man rooms talking to the other three helo pilots of the *Midway* unit.

The admiral arrived on the carrier about 1750 hours, and we were back aboard the *Helena* by 1800 hours.

The war of rhetoric goes on. Secretary of State John Foster Dulles declared today that the defense of Quemoy was the same principle as the famous airlift of West Berlin during the Red blockade in 1948 and 1949. He said the Quemoy situation is not just some island of real estate, but rather a Chinese and Russian Communist challenge to the basic principles of peace, and that armed force should not be used for aggression.

Friday, September 26

We had another day fraught with dangerous air ops, but no fatalities. We are still on the edge of the typhoon. The seas were moderate, but there was a heavy ground swell that caused all the ships in the taskforce to roll and pitch violently.

An F8U *Crusader* was trying to get aboard the *Midway* this morning. Just as he started to snag a wire, the stern rose up extra high. The landing gear of the fighter was sheared off and the plane caught fire.

The pilot added power, went into afterburner as he banged the tail of the plane against the flight deck and by using brute force, he climbed out to gain altitude.

He was streaming hydraulic fuel from the ruptured oleos. He managed to reach 5500 feet before his controls turned to mush as he lost all hydraulic fluid. The plane nosed over and pitched down just as he ejected. The plane screamed vertically down and crashed into the sea ahead of the *Midway*. A *Midway* planeguard helo piloted by Ensign Sam Anderson snagged the pilot for the rescue—number 671 for HU-1.

An hour later an A3D "Whale" jet bomber crashed on the flight deck as a result of the extreme pitching. There were no serious injuries, but there was a badly banged-up plane.

Meanwhile on the *Columbus*, Lieutenant jg Chuck Fries fought the roll on the cruiser flight deck as he was trying to land. He pounded his tailwheel hard on the deck and jammed it into the fuselage pylon, breaking off the wheel and its strut. His aircraft is now out of commission until major structural repairs can be made at Oppama.

I got in 1.9 hours and seven shipboard landings without incident. I was lucky—real lucky. These were the worst landing conditions yet.

The crew said they saw the admiral standing back by the 8-inch gun turret intrigued as he watched my touchy landings. He didn't fly anywhere with me today because the swells were so severe.

I took my 8mm movie camera on the flight deck, and when it was level, I pointed the lens toward the gray cloudy horizon. The roll would have the lens pointing about 30 degrees above the horizon for about 7 or 8 seconds, and then follow through the arc of the horizon and for 7 or 8 seconds point into the waterway below the horizon. I noticed several of the seasick ship's company barfing over the rail.

We should reach Okinawa tomorrow. We're scheduled to anchor there Saturday and Sunday. This is in preparation for the change of command ceremonies as Vice Admiral Beakley is relieved by Vice Admiral Kivette aboard the *Midway*.

Tonight Beakley ate a farewell dinner with the senior officers in the wardroom. I was placed at the head table and sat across from him. We talked and laughed a little, and he jokingly punched me in my full stomach as he left with the comment, "Keep that helicopter flying."

Monday, September 29

Ever since Admiral Beakley became COMSEVENTHFLT, the HU-1 unit assigned to him kept a little replica of the South Sea Islands NOA ceremonial mask hanging inside the helo as a memento. It was a tradition. The admiral's nickname is "The Beak," and the hand-carved wooden mask had a long Pinocchio-type nose.

We had our own farewell ceremony on the fantail, and I presented him with the humorous mask. He thought it was fun and asked the photographer to take some pictures. "The Beak" must be short for his name because he didn't resemble the mask.

As we stood around the helo talking, he expressed his appreciation for the safe flights we've had for the past few months. He volunteered to write me a letter of recommendation when I got out of the Navy, if I wanted. Of course I accepted his offer.

Tomorrow I fly Beakley out to the *Midway* for the change of command and fly Kivette back as the new commander of the Seventh Fleet. We've got our helo spit polished, waxed, and shining brighter than a new car. It will contrast with the dirty workhorse planeguard helos on the carrier.

Tuesday, September 30

The *Bennington* joined the taskforce for the change of command. They had four helos. The *Midway* had two; including mine, we had seven HU-1 helos airborne at the same time. Quite a feat and probably matched any up status at the squadron on any given day.

The seven of us air taxied the senior commanders from the taskforce ships to the *Midway* for the ceremony.

During the change-of-command ceremonies Vice Admiral Beakley, with his thoughts dwelling on the Quemoy-Matsu situation, said he wished his armada included an old-fashioned battleship.

"I have high regard for the battleship," he told newspeople.

After the ceremony, the new COMSEVENTHFLT, Vice Admiral Frederick N. Kivette, said "as long as communism is in the world we will have troubles."

Nearly all of the 3400-man crew of the *Midway* stood on its flight deck in dress whites this morning for the tradition-steeped change of command while the ship cruised on alert through Formosan waters.

Kivette's flag was run up after Beakley's was lowered. Beakley left the carrier shortly after the ceremony by plane for Manila. He will fly to Washington to become deputy chief of Naval Operations for Plans and Readiness. (He was later to join the Venerable Order of The Gray Eagle as the naval aviator on active duty with the earliest date of designation as a naval aviator.)

Kivette had served with the Seventh Fleet from 1954 to 1955 as commander of the Taiwan Patrol. The 1955 evacuation of the Nationalist Chinese garrison and civilian population from the Communist-threatened Tachen Islands occurred under his command. He was no stranger to the area. The Nationalists have said they will never give up the Matsu and Quemoy offshore islands as they did the Tachens. Kivette returned to the Seventh Fleet from Washington, where he most recently was assistant chief of Naval Operations for Air.

The new admiral seems genuinely friendly. In contrast to Beakley, he wore a helmet with earphones so that he could talk to me on the intercom. He will be fun to fly with.

A record was set with a total of 59.1 hours and 171 shipboard landings this month.

The tension remains. It was the 39th day of shelling and blockade of Quemoy and Matsu. The nationalist sea convoys, aided by some of our ships, made four landings of supplies. There are also daily air drops on Quemoy.

Yesterday an LSD that the United States had given to the Nationalists reached Quemoy with a large shipment of 8-inch Howitzers. This was the biggest artillery ever placed on Quemoy.

The American Marines moved the deadly accurate guns into firing positions, and then turned them over to Nationalist artillerymen to counter Red China's massive saturation shellings of the Quemoy Island complex. The U.S. Leathernecks escaped injury when suspicious Red gunners fired their radar-controlled artillery at the new Howitzers.

The 8-inchers—more accurate than the 105- and 155-millimeter artillery on Quemoy—are believed to be partly responsible for the recent Nationalist destruction claims against Communist artillery, ammunition dumps, and China Coast shipping.

In addition, we gave the Nationalists some cargo aircraft, C-119 Flying Box Cars, that have a 5-ton cargo capacity—twice that of the airplanes they've been using. They will be used to parachute supplies to the beleaguered Quemoy. Our government is helping them increase their capability.

It was revealed today, for the first time, that the new heat-seeking U.S. sidewinder air-to-air missiles are being used against the Red MiG-17s. That's evidently why the Nationalist Air Force has made so many recent kills.

Thursday, October 2

I ferried Admiral Kivette and Rear Admiral Southerland from the *Lexington* today. Admiral Southerland is similar to Admiral Kivette—he's friendly, always has a smile for everyone he comes in contact with, and just handles himself in a way that commands respect.

Saturday, October 4

I had a rough time today. Another battle with the bridge. The admiral stayed overnight on the carrier. The weather continues to make operating conditions difficult. I was supposed to go pick up the admiral, but the ship would not turn out of formation to give me the proper wind and deck roll conditions, so I refused to fly.

Word came back that "the old man" (captain) was blowing his stack up on the bridge. He decided to have the carrier helo pilot bring the admiral back. In spite of their big and relatively stable flight deck, the carrier helo pilots also refused to fly in the typhoon-generated weather.

But the admiral wanted back on his flagship. So after a few messages passed back and forth between the ships, the *Helena* gave me the conditions I needed. I retrieved the admiral.

Sunday, October 5

The ship pulled into Subic Bay in the Philippines and tied up to a pier. We're in the land of San Miguel beer and shiny brass belt buckles. We off-loaded and will spend 7 days at Cubi Point Naval Air Station.

The dents in the hull of the ship are scheduled to be repaired. Meanwhile we set up business in hangar space for the HUP. We also have office space, telephones, and a cumshawed panel truck. With the units from the *Midway* and *Columbus* also in this busy port, we have six HUPs and seven pilots from HU-1 units. It's like old home week being with a bunch of pilots from the squadron. The BOQ is the greatest, almost new, and like a luxury hotel compared to the ship's hot and humid stateroom. There are only two men to a room, four men to share a bath, and an individual wash basin in each room. Wow!

The HU-1 crew from the *Midway*, some reconnaissance photo guys from VFP-61—who are small detachments aboard the carriers—and I had a great picnic. We played baseball and volleyball and shared a lot of good food. Since we're so close to the equator, it was really hot. I was afraid to go swimming in the bay because of the stories about coral poisoning and barracuda swimming in the area. We're all stiff tonight from the exercise.

There's not much VIP flying since all the VIPs are together here at the air station. So I have some free time for myself.

Tomorrow I plan to take some of my crewmen flying down to Sangley Point Naval Air Station in Manila. We'll stay overnight and visit the city of Manila and fly back the next day.

One of the helo pilots from the *Midway* got word tonight that his father had died. They were evidently very close, and he was feeling very low. He should get a hop out of here for the States tomorrow. He'll be on emergency leave a week or 10 days. I felt badly for the poor guy. I really didn't know what to say. What can you say when a guy loses his pal and father?

Eisenhower is trying to pour some oil on the rough waters of the conflict in the Formosa Strait. He said at a news conference today there needs to be a cease-fire to create "an opportunity to negotiate in good faith." He must be ready to pull back our forces if he can find a face-saving way to do it.

Monday, October 6

Maybe the once-escalating activities in the Formosa Strait are slowing down. The Reds have announced that because "of humanitarian considerations" its bombardment of Quemoy, which has been going on daily since August 23rd, will cease—if there's no American escort of convoys to the island.

The Chinese Defense Minister Peng Tchhuai said the Chinese had no quarrel with the United States and that the possession of the islands was an internal Chinese matter which should be settled through negotiations.

We'll see what the United States says.

The Nationalist Chinese government responded with a skeptical cease-fire as long as the Reds observe the truce. They also stated they didn't want any direct negotiations with the Reds because the Reds never kept their word.

I flew the 50 miles to Manila and landed at Sangley Naval Air Station. Then I took a Navy boat with about 55 other guys from the base across Manila Bay to the City of Manila. Walking around town I found it to be a dirty city. There were scattered houses made from sheets of corrugated iron slapped together; raw sewage running in ditches; and dirty, dusty roads, and yet some of the young girls were dressed in white dresses walking in the middle of such filth. The most memorable thing about the place are the many movie theaters with large lighted marquees similar to those in Times Square. The sides of all the large buildings are painted like huge billboards. It reminded me a little of Las Vegas—only dirty.

Tuesday, October 7

I flew back to Cubi Point this morning. I passed by Corregidor Island at the entrance to Manila Bay. It was on that island that the famous World War II Bataan Death March started in 1942. I reflected on the fate of those men and how important it is to have a strong defense—especially now with the threat of the Red Chinese.

Just before arriving at Cubi, one of the *Midway*'s HUPs had an engine failure over Subic Bay. The magnetic chip detector warning light came on for a few seconds before the engine quit. The pilot was Lieutenant jg Sam Anderson. He was able to do a quick stop autorotation since he was cruising at 100 feet and 60 knots. He dropped the helo into the water on its belly and rolled it over on its side to stop the blades from churning. He and his crewman climbed out the cockpit window.

The engines on these machines leave a lot to be desired. If they ever get a pure jet turbine engine to fit in a helicopter, it ought to provide better reliability.

The jets from the *Midway* are operating from the Cubi Point airstrip while the ship is tied to the pier. Tonight they're all getting in some night flying and doing bounces. That includes the F8U *Crusaders*, F3H-2N Demons, FJ-4 Furys, AD Skyraiders, and A3D Sky Warrior jet bombers.

The Formosa Strait truce was broken briefly today when the nationalists fired on eight Red planes flying over Quemoy. No kills were reported. They obviously couldn't resist the temptation.

Our government said today that it will stop Seventh Fleet escort activity and would resume only "if the Chinese Communists attacks are resumed."

Our intel guys reported publicly that the Reds are busy installing new, bigger, and more powerful 12-inch Russian artillery. That would put Quemoy and Matsu easily in their range.

I wonder what effect this will have on our intense fleet operations of the last several months?

As a warning to the Reds, CNO (Chief of Naval Operations) Admiral Arleigh Burke said Red China must not be permitted to "maneuver us into a position where we are willing to sacrifice our liberty—or somebody else's." He said, "Quemoy could resist a Red invasion for 10 years—or even forever."

Soviet Premier Khrushchev is backing down. He stated the United States had incorrectly interpreted what he had said. He said his previous warning did not "contain the slightest hint" that Russia was "ready to take part in a civil war in China."

So it appears that the war of bullets is turning into a war of words as both sides verbally back down from the boiling point.

I have enjoyed the break from the shipboard activities here at Cubi. I've stayed away from the ship. The air station living is the greatest, and the food is excellent.

The *Midway* pulls out tomorrow. The helo crew has gone back aboard the ship. I don't know what they're going to do for a second helo since the loss of the one Sam Anderson dropped in Subic Bay. Having only one helo will put real pressure on the mechanics to keep it in an up condition all day, every day for planeguard duty.

Chuck Fries from the *Columbus* and I flew back down to Manila in my helo today for a sight-seeing trip. His crew is still working to repair the tailwheel damage caused during the rough seas.

I flew to the U.S. Embassy in Manila and made an approach and landing on the helo pad to familiarize myself in case the admiral wants to go there someday.

Being the commander of the Seventh Fleet's personal pilot gives me great license to fly wherever I want without a lot of questions being asked.

Chuck's ship goes to sea Friday morning, and the *Helena* leaves Saturday afternoon. The party's almost over.

CHAPTER THIRTEEN

Hoi Wong

Thursday, October 9

"DAN, YOU BETTER get your butt in gear," Chuck Fries told me on the phone. "The *Helena* is on its way back to Subic Bay to pick you and your crew up at 1600."

I glanced at my wristwatch. It was 1530 hours—a half hour before the ship arrives. I'd been out playing baseball at a picnic and just arrived in my BOQ room tired and sweaty.

Chuck didn't know the reason for the emergency. Things in the Formosa Strait had been fairly calm. He said he was also being ordered to bring his helo unit aboard the *Helena*.

It will be very crowded, I thought, with two helos jammed on the small flight deck.

I threw all my belongings into my parachute bag. I hadn't expected to go back aboard the ship until Saturday. Just this morning the *Midway* helo crew left with the carrier. The *Helena* also left port early in the morning for several days at sea for some exercises and gunnery drills. As it left the harbor, I flew two radar calibration hops for a total of 3.4 hours. The idea was to give them a spot to ensure that the guns and the radar were aligned to shoot in the same direction.

Then I attempted to secure my crew early.

I got on the phone to our hangar office and got ahold of Pobst. He said Karasinski was there with him, but they didn't know where Dunlap and Wilson were.

"Get the helo ready to go," I ordered.

"Get your gear together. We leave in half an hour."

"Yes, Sir," he choked.

The other two crewmen were on liberty. It was anybody's guess where they were. But my guess was they were in Olongapo, the nearby resort city which featured cockfights, jeepneys, hustlers, bumpy roads, women, and San Miguel beer.

I contacted the Armed Forces radio stations and had them make repeated announcements for the two crewmen to return. I also had the Armed Forces Police scour the area in search of my missing crew.

Using the name of the commander of the Seventh Fleet sure got a lot of action, but no results.

Then I made a mad dash to the hangar.

Chuck had rounded up all four of his crewmen. They and my two crewmen loaded our gear, parts, and tools on the helos.

I flew most of our gear out to the *Helena* as it entered port. It took five trips ferrying back and forth. The last one was in the dark. So far all night landings have remained the same—scary.

After my first trip, Chuck flew his helo aboard. The crew folded the blades and stored the helo aft in a small space on a narrow strip of deck between the guardrails on the starboard side of the ship and the hangar hatch for the Regulus missile.

That left the flight deck clear for my helo.

Chuck and I agreed that since this is my ship, my helo would occupy the flight deck and his would be the helo that was stowed away and would operate as backup.

After getting the helos aboard and tied down, with blades folded and stored for the night, we found out the reason for the rush.

A Norwegian freighter, the *Hoi Wong*, with mostly Chinese passengers had gone aground on October 6th, about 4 days ago, on Bombay Reef in the Paracel Islands located about 400 miles south of Hong Kong and 200 miles east of Turan, Indochina. The ship was on its way between Swatow in Kwangtung Province in Red China and Singapore. It had to make only one basic turn to navigate that nearly 2000-mile distance, but the captain had turned too soon and struck the unmarked reef at 0230 hours.

The number 2 hold was already flooded and the hull was leaking. The master of the ship feared for the safety of his passengers and had requested assistance from the Seventh Fleet through the U.S. Naval Liaison Office in Hong Kong.

The message to us read:

FM ALUSLO HONG KONG 090251Z

TO COMSEVENTHFLT
NORWEGIAN CARGO VESSEL HOI WONG WITH 130 DECK PASSENGERS SINGAPORE CHINESE ENROUTE SWATOW TO SIN-GAPORE HARD AGROUND 060240H NORTHEAST SIDE BOMBAY REEF PARCEL ISLANDS X MASTER REPORTS 15-20 FT SWELLS BOTH SIDES, IMPOSSIBLE TO APPROACH FROM LAGOON X NUMBER TWO HOLD LEAKING X MASTER CONCERNED SAFETY LIVES OF PASSENGERS AND CREW X REQUESTS HELICOPTERS UNDER PRESENT CONDITIONS

There had been at least two attempts to refloat the vessel during high tides by two tugs from Hong Kong. The efforts were unsuccessful.

Another cargo vessel, standing by, attempted to transfer the passengers into lifeboats. He was able to get one boatload transferred, but the little lifeboat got so banged up by the 15- to 20-foot swells, it was considered too dangerous to resume the operation. So there everyone sat and the request went out for help.

The irony of fate was good Samaritan efforts of the United States—we would now attempt to rescue relatives of over 100 Communist Chinese with whom we were ready to go to war just a few weeks ago.

The entire ship vibrated all night as we sped at flank speed to arrive on the scene of the grounding by 0930 hours.

Chuck and I and our crewmen didn't know what to expect. We were to be the last resort. The *Helena* was planning the rescue effort by using the ship's lifeboats. Trying to rescue that many people by helicopter, two or three at a time, was considered too hazardous and a risk fraught with danger.

Friday, October 10

We arrived at Bombay Reef on schedule and sat in deep water about 2 miles from the *Hoi Wong*.

Admiral Kivette wanted to get an up-close look to evaluate the situation. This was his show. He could have sent some other ship to the rescue, but he chose his flagship to do the job. Once again the captain was taking a backseat to events on his own ship. Usually the captain of any ship would bask in the limelight of what was going on, but not aboard the *Helena*.

The crew got my helo ready to launch. The admiral climbed aboard, and we were off for a close-up tour of the ship in distress.

The seas looked deceptively calm. We circled around several times and hovered alongside the ship. The boilers had been shut down, and there was no power on the ship.

I tried to figure out a place to hover for the rescue operation if the helo was used. There were masts, cargo booms, and rigging all over the rusty ship. The only place possible to pick up anyone from the ship was on the very front of the bow, but there was still only enough clearance for the rotor blade tips if we barely hovered half the helo over the bow. Some of the rigging would have to be cut down to allow us to hover not to endanger those under us and eliminate the possibility of clipping off the rotor blade tips.

The good thing was the wind was blowing from the stern of the ship so the back of the helo could hang over the bow as rescuees lined up to be hoisted off. This way we would have max power available for lifting rescuees as opposed to using it up in a downwind hover.

After the admiral felt he'd seen everything he wanted to look at, we flew back to the ship. The admiral and his staff huddled to plan our next move.

I returned with a couple of photographers so they could capture the situation on film from different angles. After they shot a variety of pictures, we came back to the *Helena* and landed. I shut down the engine, and the helo was secured.

The ship's lifeboats were launched and tried to get near the *Hoi Wong*. About three-quarters of the *Hoi Wong* was stuck on the reef with waves washing over the reef and undulating alongside the ship.

There was tremendous wave action on the stern of the ship because the wind was pushing against the stern. That meant the rescue boat had to be in reverse and would be bobbing like a cork while the passengers tried to slide down a rope or crawl down a rope ladder to get into it. The wooden Jacob's ladder alongside the ship was so far forward that it was over the reef and useless.

While flying the admiral, I took a look at the passengers. They appeared too frail to try shimmering down a rope like a stuntperson.

Finally one of *Helena's* boats got banged up against the ship from the rough seas. The boat crew determined that it was too dangerous for them to try to bring anyone back to the cruiser.

At that point we got word that the helos would be used. "The last resort was on."

So we could carry additional weight, we removed the forward passenger seat, instead of just folding it forward and out of the way of the rescue hatch. We also removed all excess gear, life rafts, smoke bombs, and radios. The helos were stripped clean. The only thing left for the crewman were the bolt cutters near the hoist. If the hoist cable became entangled, the crewman could then quickly cut the cable, and we would be free from being tied to the ship.

We all had a concern about hoisting the frail women and small children in the rescue sling. We figured the men would be okay.

Like we learned in survival school, if you put the sling on backward, you would slide out. We had visions of somebody getting in backward or being too small and sliding out and slamming onto the deck of the *Hoi Wong* or hitting the edge and falling into the sea.

The new three-prong rescue seat being developed back in the squadron would have been perfect for this event. But it hadn't been approved for use in the fleet yet. However, ingenuity prevailed. The *Helena's* crew sewed together four canvas buckets. Each had two holes in the bottom and a large enough circumference at the top for a passenger to fit inside wearing a kapok life jacket. It was like a giant pair of underpants.

On top it had a 1-inch spliced rope with an eyehole that resembled a breeches buoy and would hook onto the helicopter rescue hoist snap. The lead ball on top of the rescue snap would take the weight down to the ship. This would replace the rescue sling and would ensure us that nobody would slip away from us while we were hoisting them into the helo. Of course it would be slower getting our rescuee out of the canvas bucket while we hovered—but safer.

The people on the *Hoi Wong* stood in line to be hoisted into the helo. Then they stepped into the canvas bucket and put their feet in the holes. Most were wearing black or dirty white-colored Chinese-type pajamas. With two buckets for each helo, somebody could be getting prepared in one bucket while the other one was being hoisted.

My next flight to the *Hoi Wong* was to deliver two officers from the ship. I lowered them on to the forecastle. Their job was to ensure that the cables and jack staff were cut and removed from the bow to allow us the room to hover. That way there would be a 15- to 20-foot clearance between the rotor tip path and the nearest ship's mast and stays. It was still tight quarters, but we had a little leeway for any mistake or shifts in the wind.

While I was on a hop, Chuck Fries moved his helo to the flight deck, spread the rotors, cranked up, and prepared to launch. He was transferring two additional men and his crew chief AD1 E. C. Magee, to the *Hoi Wong*. The crew chief would take responsibility for fitting all the passengers securely and safely in the canvas bucket or ensuring adults were properly fitted into the rescue sling.

For more than 4 hours, there would be room on the cruiser flight deck for only one helo at a time. If one of us had a problem or an emergency, the other would have to be prepared to lift off immediately. There was no alternative location to land except in the sea. If we landed on the reef itself, we'd probably get torn apart by the sharp jagged coral and no boat could get near for a rescue.

We refueled on the run, that is, with the engine and rotors still turning. We were very glad the gasoline tanks had been completely drained, cleaned, and tested while the ship was in the Subic shipyard. No more slow refueling with our chamois bucket.

The *Helena*'s crew lined the rails and superstructure to watch the operation. A lifeboat was launched and drifted in the water near the freighter to rescue us in case there was a problem with either of the helos.

While Chuck was pushing his helo from the aft storage area to the flight deck, preparing his helo for flight, I hoisted the first people aboard my bird.

All the rescuees were terrified by the helicopter. This was probably the first time they'd ever seen one, much less ever ridden in one.

They would cry, hold onto the interior frame of the helo with white knuckles, scream, and wet their pants during the short flight. It was wild in the back of the helo. We were careful to ensure they all faced aft so that they wouldn't grab the idle mixture cutoff handle.

None of the Chinese spoke English, so everything on the ship deck was done with hand signals. Only the crew and those to be hoisted were to be directly under the helo. All others stood down the deck in a safe area.

I considered hoisting only two on my first hover, but when I found I still had plenty of boost, I hoisted two more. These Chinese were so small and skinny, they weighed practically nothing. They brought a few of their precious belongings with them, but most of their worldly possessions were left behind on the *Hoi Wong* until it rusted to pieces or broke apart from the heavy seas.

On future hoists our limit was not measured by the number of people, but by power left to maneuver the helo. The temperature was 86°F (86 degrees Fahrenheit) with a lot of humidity, which made for a high-density altitude. That restricted our lifting ability. On one trip I hoisted nine women and children into the helo before I used maximum power—a record for the most people ever hoisted into the HUP, and a record for the most people ever flown in a HUP—11 including pilot and crewman.

Because the radios were stripped from the helos, there was no communication with the ship or between helicopters or between us and the *Hoi Wong*—just the intercom between the pilot and the crewmen—and we were too busy for any idle chatter.

It took Chuck and me more than 4 hours to transfer everyone to the *Helena*. We rescued a total of 106 persons in 27 rescue trips. I made 14 trips carrying 58 people, and Chuck made 13 trips carrying 48 people. A total of 44 women, 22 children, and 40 men, including one Englishman, were rescued.

Including the crew transfers, I completed a total of 21 cruiser landings during a stretch just over 4 hours. We finished the project at 1630 hours. We were all very tired, but happy.

The admiral greeted each of the mystified Chinese as the helos landed them on the *Hoi Wong*. Fear was still mirrored on their faces. I don't think they knew who he was, but surely noticed all the fancy shoulder markings of gold braid on his white uniform and the scrambled eggs on his hat. Nevertheless he extended the warm hand of American friendship as he ducked under the rotor blades. The Chinese were so short, they could fully stand up without fear of being decapitated.

The body odor of the Chinese survivors was terrible. On one flight I tried to fly with my head out the side window, the smell was so foul. It didn't take the ship's crew long to strip them from their clothes. The ship's laundry washed their clothes while all the rescuees were instructed to shower down.

The helicopter crewmen rotated during the day. Chuck and I never got unstrapped.

After we rescued all the passengers, we retrieved our crewmen and officers from the *Helena*. A few men from the cargo ship's crew elected to stay aboard for a few more days to see if they could save their ship. They were all able-bodied men. The ship was rammed hard on the reef, and from my vantage point I was convinced it would remain there forever. The two tugs remained idling nearby. They were safe enough unless another typhoon came along.

The captain had nowhere to go. Since a person never gets a second chance to run a ship on a reef, once he left his ship, the captain's seaborne days and command as a ship's master were over. "So what's the rush," I thought.

After the rescuees were cleaned up, they were treated to plenty of food, capped off by lots of ice cream and candy. The American hospitality was working overtime. Most of them were too shy or frightened to do much except sit around, do what they were told, and enjoy the hospitality lavished on them.

Mickey Mouse movies were shown to the kids. The Seventh Fleet band performed in concert for our overwhelmed guests.

Soon all the passengers started to get sick. The ice cream was just too rich for people who'd always existed on a simple diet of white rice. What a mess!

We flew out walkie-talkies and food supplies to the tugs. The tugs hadn't realized they'd be staying at sea as long as it now appeared they would.

I logged a tiring 4.3 hours, most of it demanding hover time in tight quarters. I was really pooped. I had no complaints about not enough flying today.

After we secured flight ops, the crew started putting our helos back together again. There was a lot of gear, including our radios piled on the edge of the flight deck. Next the crew gave each helo a thorough freshwater wash to get rid of all the urine and excrement left behind by our frightened passengers. Of course, there were plenty of jokes about how flying with us would scare anyone—but that really wasn't the truth.

Chuck and I got a call from Admiral Kivette to go to his cabin. He was effusively friendly and invited us to sit down. He told us that he thought we'd done a great job.

"I'd like to hear about how it went today," he asked. "Give me the details."

We tried to share our individual experiences and observations of the day's activities.

"You men must be exhausted. Do you need a medical prescription?" he asked.

After a few minutes it dawned on me that he was trying to tell us that he was going to have the ship's doctor give us a couple of bottles of whiskey.

"That's alright, Sir, I'm okay. I don't need anything," I responded as a nondrinker.

He insisted, and Chuck was only too happy to oblige. I kept my two bottles as souvenirs.

We sat around sipping ice tea just shooting the bull with him. This was heady stuff for a couple of junior jg's. He said he was going to send a message to the squadron praising our actions. Some of the staff were talking about a medal, but I doubt that—the whiskey from the doctor probably will be our best reward.

Walking around the ship tonight is quite an experience. A lot of the crew are walking up asking questions, wanting pictures together, and generally carrying on about the day's rescue activities. It feels sort of embarrassing to just do what you love to do and end up being a junior-grade celebrity.

The ship's now heading for Hong Kong to deliver our rescued passengers. I'm trying to figure out how to get my crewmen from Cubi Point back aboard the *Helena* since we're not returning to the Philippines for a month or two. I also owe money for the gas I used while there and left some gear behind in our sudden departure.

There were several congratulatory messages sent out by the admiral.

CONFIDENTIAL

P   100932Z

FM   COMMANDER SEVENTH FLEET

INFO   CHIEF OF NAVAL OPERATIONS - WASH D.C.

TO   COMMANDER IN CHIEF PACIFIC FLEET

PERSONAL FOR ADMS BURKE, FELT AND HOPWOOD
I HAVE JUST UNCROSSED MY FINGERS WITH THE LAST CHOPPER LANDING X EXCEPT FOR THE CHOPPER PILOTS THIS WAS THE EASY WAY TO DO IT X BECAUSE IT WAS EASY DOES NOT MEAN THAT IT WAS NOT A SPLENDID ACCOMPLISHMENT AND A DRAMATIC PERFORMANCE BY HELENA X I WOULD HAVE PREFERRED TO TRANSFER THE RESCUED PASSENGERS TO THE OTHER NORWEGIAN SHIP STANDING BY BUT THE SWELL WAS STILL RUNNING HIGH AND THE WIND INCREASING X THE MAJORITY OF THE PASSENGERS WERE WOMEN AND CHILDREN AND I THOUGHT IT WISE NOT TO CROWD MY LUCK X OFFICERS AND CREW REMAINED ON THE HOI WONG WHICH IS IN NO DANGER SO LONG AS THE WEATHER REMAINS MODERATE X MOST PASSENGERS WERE UNNERVED BY BEING HOISTED INTO HELICOPTER BY RESCUE GEAR AND FLIGHT TO HELENA BUT ARE IN GOOD SPIRITS NOW X CREW REACTION TYPICAL AMERICAN FRIENDLY CURIOUS GENEROSITY X IF I CANT UNLOAD THESE PEOPLE TOMORROW I WILL PUT THEM ON THE STATION SHIP X I STILL FEEL LIKE A MEDICINE MAN

R   101212Z   UNCLASSIFIED

FM   COMSEVENTHFLT

TO   HUTRON I

INFO   HUTRON I DET I
         COMAIRPAC

   TODAY YOUR HELICOPTER UNITS 6 AND 7 OPERATING
FROM NORWEGIAN VESSEL HOI WONG, AGROUND ON REEF
PARCEL ISLANDS X ROUGH SURF AND SHALLOW WATER PRE-
VENTED ANY OTHER METHOD OF RESCUE X EACH HELO OPER-
ATED CONTINUOUSLY FOR MORE THAN 4 HOURS CARRYING
ON SOME SHUTTLES UP TO 9 WOMEN AND CHILDREN PAS-
SENGERS AT A TIME X OPERATION WAS CONDUCTED IN
CLOSE QUARTERS FROM FORECASTLE OF STRANDED SHIP BY
HOIST X LTJG FRIES AND MCKINNON DID AN OUTSTANDING
JOB X THE RESOURCEFULNESS AND FLIGHT PROFICIENCY DIS-
PLAYED BY BOTH OF THESE YOUNG PILOTS REFLECTS CREDIT
ON THEIR PARENT ORGANIZATION X AM COMMENDING
BOTH OFFICERS AND THEIR CREWS BY SEPARATE CORRE-
SPONDENCE X
                                                    VADM KIVETTE

FM   CNO

TO   HUTRON I
   THE OUTSTANDING PERFORMANCE OF HELENA AND
EMBARKED HELICOPTER DETACHMENT IN RESCUE OPERA-
TIONS AT PARACEL ISLAND IS AN INSPIRING EXAMPLE OF
THE READINESS OF THE U.S. NAVY TO RENDER HUMANITAR-
IAN ASSISTANCE WHEREVER NECESSARY X ACTIONS SPEAK
LOUDER THAN WORDS AND YOURS WILL SERVE TO HIGH-
LIGHT TO THE WORLD THE SELFLESS DEVOTION TO DUTY OF
THE OFFICERS AND MEN OF THE U.S. NAVY IN ALL JUST AND
HUMANITARIAN CAUSES X CONGRATULATIONS FOR YOUR
SPLENDID ACHIEVEMENT X TO ALL HANDS, "WELL DONE X"
                                                    S/ Arleigh Burke

FM   SEC NAV

TO   USS HELENA
   MY CONGRATULATIONS TO YOU AND THE OFFICERS AND
MEN UNDER YOUR COMMAND FOR THE SUCCESSFUL RESCUE
OF SURVIVORS FROM THE MERCHANT VESSEL HOI WONG X
SUCH HUMANITARIAN EFFORTS DESERVE THE THANKS AND
PRAISE OF ALL WHO TRAVEL THE SEA LANES OF THE WORLD
X IT IS PARTICULARLY GRATIFYING TO SEE SUCH HEART-
WARMING RESCUE OPERATIONS IN AN AREA TROUBLED
WITH INTERNATIONAL MISUNDERSTANDING X
                                                    Thomas S. Gates, Jr.

FM   CINCPACFLT

TO   USS HELENA
    THE RESCUE OF 106 PASSENGERS FROM THE STRICKEN SHIP,
HOI WONG, HAS MADE AN OUTSTANDING CONTRIBUTION
TOWARD STRENGTHENING OUR RELATIONS WITH ALL ASIAN
NATIONS AND WILL BE NOTED BY OUR FRIENDS THROUGH-
OUT THE WORLD X YOUR SKILL AND COURAGE IN CARRYING
OUT A DIFFICULT HUMANITARIAN MISSION WITH EFFICIENCY
AND COMPLETE SAFETY IS ANOTHER EXAMPLE OF THE VERSA-
TILITY OF SEVENTH FLEET UNITS X SPECIAL KUDOS TO THE
HELICOPTER PILOTS AND CREWS FOR A MAGNIFICENT JOB X
TO THE OFFICERS AND MEN OF THE HELENA AND THE HELI-
COPTER DETACHMENT THE HEARTIEST "WELL DONE" X
                               /S/ Adm. H.G. Hopwood

M   100910Z   UNCLASSIFIED

FM   COMSEVENTHFLT

TO   USS HELENA

INFO   CNO
      CINCPACFLT
      COMCRUDESPAC
      COMCRUDIV 3
      Commander Cruiser Division 3

    FOLLOWING RECEIVED BY COMSEVENTHFLT FROM MASTER
STRANDED MERCHANT SHIP HOI WONG QUOTE THANKS TO
YOU AND YOUR MEN FOR THAT WONDERFUL JOB YOU DID
TODAY STOP IT CERTAINLY IS A GREAT RELIEF FOR ME THANKS
AGAIN UNQUOTE AND I AM PLEASED TO ADD I COULD NOT
AGREE MORE X MY CONGRATULATIONS ON A JOB DONE IN AN
OUTSTANDING MANNER X
                               VADM KIVETTE

Meanwhile back in the Formosa Strait there was a dogfight over
Matsu Island between the Nationalists and the Reds. Five Red MiGs and
one Nationalist F-86 were shot down in the clash, but the truce in the
Quemoy area is still holding.

President Chiang Kai-Shek rubbed the Red Chinese noses in the mud
when he asserted today that the Nationalist troops had "won the first
round in the battle of Quemoy and would eventually deliver our compatriots on the mainland to freedom from communist tyranny."

He went on to say that the Communists had been put into "a hopeless
position from which they can never extricate themselves." He challenged

the Chinese Reds to attack again and "invited common action by the free world's antiaggressive force."

I'm sure the guys in Washington aren't too hot to get into a war to free the Chinese mainland that Chiang lost in 1949. What they seem to want to do is contain Communist aggression.

Saturday, October 11

After steaming all night we arrived in Hong Kong late this morning. The ship is to stay 4 hours while transferring our Chinese guests ashore. After conducting a news conference, we head back out to sea.

This is a beautiful, busy harbor with picturesque Chinese sailing junks, crisscrossing ferry boats tooting their horns, sampans, and nesting junks surrounded by pockets of tall buildings and slum housing.

From a distance it looks so beautiful. A look through binoculars shows how picturesque the harbor city is, but one can also see a lot of unbelievable poverty.

As we moored to a giant buoy, an old lady with a fleet of sampans manned by other old women began to scrub the grease, oil, soot, and dirt off the sides of our warship.

"Who's that?" I asked one of the boatswain mates.

"Sally Soo and her sidecleaners."

"Sally Soo and her sidecleaners?" I repeated with a puzzled expression. "What does she do?"

"It's a tradition. She cleans off the sides of U.S. Navy ships anchoring or mooring in Hong Kong in exchange for the ship's garbage," he answered.

"What does she do with the garbage?"

"Sells all the slop as food to some of the poorer people and refugees from Red China."

My stomach almost retched as I said a silent thank-you prayer for being an American.

In later years she demanded cash and painted the sides if the ship provided the paint.

We unloaded our passengers to a waiting ferryboat via an accommodation ladder lowered alongside the cruiser. They all looked crisp and clean. I don't think those Chinese ever had such spotless clothes. A boatload of newspeople pulled alongside, and the admiral conducted a news conference. He invited Chuck and me to attend. His invitation made it mandatory.

There were about 15 English-speaking reporters, each asking different questions all at the same time. Every news organization imaginable was there—UPI (United Press International), Associated Press, Time-Life, Reuters, *New York Times*, and others I'd never heard of. I found out

I didn't know all of what went on in the rescue effort because I was too busy flying.

The news conference lasted about 20 minutes. Some pictures were snapped, and then we took a longing gaze from the ship at Hong Kong one last time. There was no liberty since we were in port only to drop off our rescuees. The word is we will return in December for a visit.

It's been nice having another pilot aboard to talk with and trade stories. Although I wonder if I'd like it all the time since I'd have to share my flying with him. Everyone is pretty selfish about wanting to get in all the flight time he can, me included.

Tomorrow we rendezvous with *Columbus*, and Chuck and his crew will fly off when the ships join up at sea. I'll be down to two crewmen, but we'll make it work.

It turns out this was the second mass rescue in HU-1s and Navy history and the most ever rescued in peacetime. The total "saves" was equivalent to the average sum of three peacetime years by HU-1.

The previous mass rescue was made by Duane Thorin, ADC/AP, on 12 January 1951. He put 118 men ashore from the Thai ship, *Parsae*, after it had run aground in hostile waters off North Korea during a snowstorm.

Thorin's helicopter was frequently subjected to enemy fire during that rescue procedure. Now a Navy Lieutenant, Thorin authored the book *Ride to Panmunjon*, after being captured in February 1952 and spending 18 months as a POW before being repatriated in Operation Big Switch.

Today the "Plan of the Day" contained a nice note. It said, "Congratulations to all hands for the efficient and seamanlike manner in which you carried out an outstanding rescue job yesterday. A special pat on the back to the helicopter pilots, their plane crews, and flight deck crew for their fine performance during the transfer operations. *Well done.*

CHAPTER FOURTEEN

Underway Operations

Sunday, October 12

THIS AFTERNOON WE JOINED UP with the Midway and Columbus. Chuck and his crew returned to their ship.

I flew 1.6 hours ferrying mail and chaplains to destroyers in our task group.

I ferried Admiral Kivette to the *Midway*. He's an aviator, but not a helicopter pilot.

He asked me how helos fly.

"Here, why don't you try the cyclic and rudder pedals," I told him as I let him try for the first time.

I kept my left hand on the collective and right hand near the cyclic because we were bouncing around like a cork in water. He wasn't smooth, but he didn't give up. I'll teach him. He's sure different from Vice Admiral Beakley. I think I'm going to enjoy being his pilot.

The rescue saga continues. More messages and clippings arrived. I'm going to get a letter of commendation from the admiral, but the official word is that there will be no medals. The Air Force dishes out medals for virtually anything. That's not the Navy's style.

The self-imposed 7-day halt in the Chicom artillery bombardment of Nationalist-held Quemoy has been unilaterally extended for 2 weeks, Peiping radio announced today. The Red Chinese asserted that the United States was maintaining forces in the Formosa area in an effort to "take a hand in our Civil War."

It warned that the Quemoy bombardment would "start at once" if the United States resumed "escorted operations in the Quemoy water area."

President Eisenhower hailed the message as "good news" and said it would afford the United States "a further opportunity to work out through negotiations a settlement to the problems in the area."

Red China reiterated that the Quemoy question was "China's internal affair and no foreigner has any right to meddle with it, including the UN."

Naturally they didn't want anyone to help the Nationalists. The Reds could easily overpower the Nationalists as long as there was no support from the free world.

Nationalist-armed forces took advantage of the first 7-day truce to reinforce Quemoy defenses and stockpile food, munitions, and equipment on the islands. Nationalist ships and aircraft traveled unescorted and were not molested by Communist shore batteries.

Vessels returning to Formosa took with them 6000 of Quemoy's 47,000 civilians, mostly aged persons and children.

If the extended cease-fire holds, we get to go to Yokosuka Saturday. Evidently we're keeping all our warships out of the Formosa Strait.

The Communists sure have messed up our proposed operating schedule, but I guess we're out here to show a strong deterrent to their aggressive actions. That's part of our mission.

Secretary of Defense Neil McElroy met today with Chiang Kai-Shek to talk about settlement of the problem. Tomorrow he visits us—wherever we are. I'm scheduled to fly him around. I'm spending the night on the carrier to be in position for the Secretary of Defense's flights. There was a nice air-conditioned room for me.

Our daily operations extended into the night. I got caught out after dark and had to make a scary night landing aboard the carrier. With no visual references and no instruments in the helo, and the carrier flight deck was pitch black; it was really hairy.

Working on my logbook, I noticed I now have over 300 helo hours in 10 months. I hope I reach 500 before I get out of the Navy the end of next year.

Monday, October 13

Admiral Kivette introduced me to Secretary of Defense Neil McElroy as the hero of the ship rescue operation. We talked a short while prior to takeoff. I also met Admiral Harry D. Felt, CINCPAC (Commander in Chief Pacific), who is the unified commander in charge of all the armed forces in the entire Pacific area—Army, Navy, Air Force, and Marines. All the brass is out here on tour of our forces.

I did a lot of flying—6.9 hours, and one hop was 6.0 hours with 31 shipboard landings. I sat strapped in the helo from 0755 until 1455 hours. It was a very busy day taxiing between ships.

Back aboard the *Helena* tonight, we're headed for Yokosuka for a 10-day, in-port visit. We've spent a lot of time at sea, and the break will be good for me and the maintenance of my helo, better known as UP-18 because of the squadron UP painted on the tail pylon and the number 18 painted on the side of the helo.

Aboard ship the crew has a really tough time doing anything more than routine maintenance, and this machine requires a lot of attention.

I hope to get some Christmas shopping done while we're in Yokosuka. With all the dunkings of the HUPs recently, the squadron safety officer sent each unit a message reminding us about a new life raft available.

"They say that experience is the best teacher, and it's probably true. Past experience has taught us that there is little time to fumble around for survival equipment when the time comes to abandon your sinking helo. Also, as anyone who has bobbed around in the Pacific Ocean supported by only a Mae West–type life jacket can tell you, it is a humbling sensation to suddenly realize that no one knows just where you are.

"This seems like a good time to remind you that life rafts are available for you to wear on your back.

"If, and when, King Neptune should decide to creep into the cockpit with you, forcing you to evacuate that left-hand seat, your trusty Goodyear–type 'yacht' will be right at your 'side'—piggyback style.

"Whether you are going to WESTPAC or just to hover over the Coronado Roads, check out a backpack and take it along. It's free—tax included."

The only thing he didn't say was how uncomfortable it was to fly with it attached to your body. I'll take my chances.

Thursday, October 16

I was scheduled to take the admiral to a conference on Okinawa today. An inspection of the helo showed that one of the brackets that holds the rotor blades to the rotor head needs replacement. The tolerances are too risky to take a chance with the admiral. I'd hate to have something happen with him aboard.

Dunlop and Wilson made a surprise entrance today. They were aboard a tanker that pulled up alongside us for refueling and were highlined to the *Helena.*

Karasinski and Pobst were anxious to share the many details of the rescue operation. Their stories solved the two missing crewmen's curiosity, but also heightened their envy in not being involved.

Before he left the Far East, Secretary of Defense Neil McElroy spoke to an audience of Nationalist Marines and told them they'd been a big help to halt "what might have been a very destructive war."

I wonder if the United States would have actually used nuclear weapons. Although it was kept secret, every carrier had four planes on 3-minute alert with nuclear bombs attached to the wings and on the belly. Marine guards with their weapons at the ready always surrounded the aircraft. I'm not sure whom they were protecting the bombs from, because it would be an assumption the ship's company and the pilots were all on the same team. The armed planes were changed every 4 days, and the nukes were changed every 3 days, which kept deck crews and ordnancemen continually busy. The nuclear weapons training school at North Island had said they expected to lose a carrier by 1965 through some type of accident with the weapons.

Saturday, October 18

The weather was so bad that the ship arrived in Yokosuka 6 hours late. We inched into the harbor fighting the storms and high winds. Just before the ship tied to the dock, I decided to risk flying across the harbor to Oppama, where *Det One* is located.

I gambled that the rotor blade bracket wouldn't break on such a short flight. The risk seemed worthwhile; otherwise I didn't know how we'd get the helo fixed.

I instructed the crew to go by truck in case the rotor head broke and I went into the drink. I decided to take the risk alone. I didn't think it was wise to take a chance with anyone else's life. Just after takeoff, the moisture and heat inside the helo from the bad weather fogged the inside of the canopy. It made it extremely difficult to see where I was going. There was no heat to clear up the canopy.

It may have been a mistake not to take along a crewman, I thought. Such caution could cost me my life. The canopy was beyond my arm's reach, even if I dared let go of the collective, to wipe away the moisture so I could see out. At least a crewman could have unstrapped and easily reached over to wipe the moisture off.

The poor weather, fog, and low clouds on the outside and no instruments on the inside made for an alarming flight across the harbor. I couldn't go back to the cruiser because I couldn't see enough to land.

Finally, my best solution was to fly sideways and look out my side window.

Sunday, October 19

The *Lex* has arrived at Yokosuka, too. The helo crew, Lieutenant M. E. Phillips, Ensign B. J. Hale, Ensign Rolf Volonte, and John Loomis have joined me at the BOQ.

We talked about the *Hoi Wong* rescue operation. No one believed I had not exceeded 40 inches of boost. Everyone believes I exceeded the engine limits with 11 people on board the HUP.

It was especially good to see my old buddy, John Loomis. We made great plans to go to Tokyo during our scheduled 10 days in port.

The *Det One* crew would do the bulk of the heavy maintenance necessary on UP-18. There were a lot of gripes about the bird, and it would take 10 days to fix them all. They'll get started the first thing in the morning.

I attended Sunday services and got revitalized.

The weather is cool, and we're wearing our blue wool uniforms.

Monday, October 20

Communist China resumed its bombardment of Nationalist-held Quemoy, Tatan, Erhtan, and Hutzu Islands off the mainland. Peiping radio broadcasts, monitored shortly before the shelling was resumed, charged that the self-imposed Red cease-fire had been broken because of U.S. Navy escort of a Nationalist convoy in Quemoy waters yesterday.

Peiping termed the renewed shelling a "measure of punishment" for the alleged U.S. Navy action. It renewed warnings that Red China considered it "absolutely impermissible for the Americans to meddle in internal Chinese affairs."

The U.S. Taiwan Defense Command denied that any American vessels entered the Quemoy waters acting as an escort for Nationalist supply ships. However, American and Nationalist authorities later conceded that a U.S. LSD had sheltered three small Nationalist supply ships and that an American destroyer had been near Quemoy. They denied that either vessel had approached the 12-mile territorial limit claimed by Red China.

Communist shore batteries hurled 11,520 shells at the offshore islands in the first $2^1/2$ hours. It looks like they're continuing today and there appears to be no letup.

Since things have heated up in the south, we're pulling out tomorrow morning. My helo is scattered in parts all over the floor of the hangar here at *Det One*. There's not enough time to put it together. It was a great helo that had come out of overhaul before we left on cruise. Now they've got to give me another helo—UP-30. It's the only one they've got in flyable condition, and that may be an exaggeration. It needs a lot of work. My crew is on liberty, and someone has to get my substitute helo ready to go. *Det One* people are stuck with the job.

The UP-30 has 2 months to go until it's ready for a major overhaul. That means it's pretty well worn out and a real dog. What a disappointment! The carrier's detachment is having problems, too. They have got an engine out of one their helos for repair work. They're having a heck of a time getting it back in by tomorrow morning. It's going to be a long night for all of *Det One* people.

I packed tonight—just finished unpacking. It's raining slightly, and there's patchy fog. Although the weather is lousy, it's cool compared to the Philippines.

Tuesday, October 21

The crew was finally rounded up and arrived aboard the ship last night. *Det One*'s men worked all night to get UP-30 ready to fly. I was up early to fly it to the ship, but I had to wait around 3 hours for it to be completely assembled. I was right—my HUP was in too many pieces to put back together so quickly.

It was a scramble rounding up the *Helena's* crew plus those of the *Lexington* and other U.S. warships in port. Armed forces radio carried announcements for all crews to return to their ships. The Shore Patrol and Japanese police helped spread the word as well.

I finally got airborne and joined the *Helena* in the harbor as it was slowly steaming out of Tokyo Bay. As soon I was landed, I was ordered to the bridge. Instead of my fore/aft military hat, I wore a new blue baseball cap I purchased at a Japanese souvenir shop. It had a set of Navy gold wings embroidered on the front with my name stitched below. All the guys on the carrier had one.

The captain and the XO wanted to know why I had a dirty gray-colored helo instead of my nice shiny blue one. They weren't impressed with the substitute helo and didn't think the admiral would be, either.

Then they noticed my cap and forbade me to wear it. Standard headwear aboard the ship is a normal Navy hat with a bill, I was informed. They allowed me a fore/aft hat but were afraid I'd start a fad on the ship with a baseball cap. Everyone in the squadron can wear a baseball cap all over the Pacific area, except those on the flagship. Thoughts of a Navy career have evaporated, if they ever really existed. The casual atmosphere among aviators was one thing, but the strict regulations that guided the blackshoes is just not my style.

My crew leader Dunlap is having hemorrhoid troubles and has had minor surgery. He needs more surgery and will be out of action for a couple of weeks. Of course, he's the recipient of all the usual jokes from his fellow crewmen about the problem.

Wednesday, October 22

We're steaming to the Formosa Strait. The shelling that started several days ago is continuing on all the Nationalist islands off the Chinese coast. It's big time trouble.

Secretary of State John Foster Dulles has arrived in Taiwan and warned the Red Chinese that the United States would resume its naval escort of Nationalist vessels to the Quemoy area if such a move became militarily necessary. He reminded the Red Chinese of their cease-fire that had started October 8th. Our ships aren't needed to help out right now because the Nationalists had a successful period during the cease-fire to reinforce their supply buildup.

The president announced that Dulles's trip was a "mission of peace" but turned out to be a "tragedy." He expressed hope that the Red Chinese would stop their shelling before it developed into a serious conflict.

There's no flying today. The crew is busy working on the helo trying to clean it. It was one of those filthy carrier helos. It will require constant work just to keep it flying. Boy, were we spoiled with our original bird!

No one knows when we'll return to Yokosuka. Presently things have quieted down on the warfront. The feeling in general is that the Commies started the firing to greet the secretary of state on his 3-day visit. It's doubtful they'll continue the bombardment.

Thursday, October 23

An agreement was announced today between Secretary of State Dulles and Chiang Kai-Shek. The Nationalists pledged not to use military force to return to the Chinese mainland. They didn't have the strength to do so, anyway, but the argument was a face saver for Chiang. It recognized the Nationalist government "considers that the restoration of freedom to its people on the mainland is its sacred mission and that the principal means of successfully achieving its mission is the implementation of nationalism, democracy, and social well being and not the use of force."

The communiqué made it clear that the United States would continue to recognize the Nationalist government as the "authentic spokesman for free China."

It asserted the recent Quemoy bombardment had served only to draw the United States and Nationalist governments "closer together." It renewed the American agreement to consider "the defense of the Quemoys, together with the Matsus, as closely related to the defense of Taiwan and the Pescadores."

Dulles made it clear to newspeople that he told Chiang that the United States would not lend military or logistic support to any Nationalist invasion of the Chinese mainland, but would continue its support of present Nationalist positions on Formosa and the offshore islands.

The United States stuck its neck out and took us to the brink of war on this entire issue. After thinking it through, the United States has come to the conclusion that supporting all these islands is not worth the risk and is trying to make every face-saving maneuver to avert a major war.

So we may head back to Yokosuka by Monday for 9 days of liberty.

The weather remains inclement, and the ship is rocking and rolling continuously from one side to the other. I wish I had straps on my bunk. The rolling is so extreme that it rolls my body from one side of the rack to the other while I'm trying to sleep.

The admiral's staff has taken a liking to me.

"We'd like you to request to be assigned to the admiral's staff," Lieutenant Commander Brown, flag secretary, told me today.

"It would be for a tour of duty of about a year and a half, and you could bring your wife out to Japan."

"It doesn't work that way," I told him.

"The squadron assigns detachments to a ship for the WESTPAC cruise."

"You forget who my boss is," he reminded me. "When he wants something, he gets it."

"But I want to get back to the States and the squadron," I rebutted.

"Well, let's talk about this more later," he concluded, trying not to get me locked into a firm position.

I'm flattered, but I'm homesick. And I'm still scheduled to be here until January or February.

Monday, October 27

The Communist Chinese shore batteries have stopped their shelling of Quemoy and the other nearby islands on even-numbered days of the month. They say they'll shell only on odd-dated days. What a weird way to run a war.

The partial truce was ordered after the Communists had terminated the two earlier self-imposed cease-fires.

The Red Chinese Defense Minister Marshall Peng Teh-hui said that he ordered the new cease-fire "so our compatriots, both military and civilian, on the islands may all get sufficient supplies to facilitate your entrenchment for a long time to come."

He went on to say that the Communist guns would "not necessarily conduct shelling on odd dates," but he urged the Nationalists to confine their supply operations to even days to avoid losses. He also warned that the truce would be suspended if U.S. Navy vessels resumed escort duties in Quemoy waters.

Since his statement, the artillery continues sporadic bombardments. The war of words is continuing. Secretary of State Dulles labeled "outlandish and uncivilized" the strategy concocted by the Communists "in their effort to save face after it became clear that the islands could not be cut off and made to wither on the vine."

He went on to say that the shelling of the islands on alternate days shows that the "killing is done for political reasons and promiscuously."

He also expressed doubt that the Chinese would engage in a level of military effort which is likely to provoke a general war.

I received my letter of commendation from Admiral Kivette today. He said a copy of it would go in my permanent service record.

UNITED STATES PACIFIC FLEET
COMMANDER SEVENTH FLEET

From: Commander SEVENTH Fleet

To: LTJG Clinton D. McKinnon, USNR, 549767/1315

Via: Commanding Officer, USS HELENA (CA75)

Subj: Letter of Commendation

1. On 10 October 1958 while attached to the USS *Helena* (CA75) as the officer in charge of Helicopter Unit 7 of Helicopter Utility Squadron ONE, I personally observed your performance during your participation in the rescue of 106 passengers from the SS HOI WONG, a commercial vessel of Norwegian registry, hard aground on BOMBAY REEF of the PARACEL ISLANDS in the SOUTH CHINA SEA.

2. During a period of more than four hours, you flew continuous rescue trips between the HOI WONG and the USS *Helena*. Your efforts coordinated with flights made by LTJG Charles L. FRIES, USN, 558931/1310, the Officer in Charge of Helicopter Unit 6 of your squadron, resulted in the rescue of 106 Chinese & English passengers from the stranded vessel. Those rescued by both helicopters in 25 round-trip flights included 36 men, 48 women and 22 children.

3. The entire helicopter rescue operation was completed without any untoward incident or injury to any personnel, civilian or Naval. The proficiency of your airmanship was repeatedly demonstrated in your ability to hover your aircraft over the restricted area on the forecastle of the HOI WONG. The close proximity of the cargo booms and standing rigging of the stranded vessel was a constant hazard to yourself and your aircraft while conducting the rescue operations with your airborne rescue equipment.

4. I take great pleasure in commending you for your outstanding performance of duty under trying conditions. Your actions were in keeping with the high tradition of the United States Navy.

5. Copies of this letter will be forwarded to the Commanding Officer, Helicopter Squadron ONE and the Chief of Naval Personnel. The Commanding Officer, USS *Helena* is requested to include a copy of this letter with your next regular report of fitness.

F. N. KIVETTE

Copy to: CO HUIBUPERS (Jacket)

We returned to Yokosuka this afternoon. I flew off the ship about 2 hours before it docked. It was raining and the visibility was quite poor again, but I managed to find my way back to Oppama okay.

I spent last night aboard the ship filling out all my month-end reports for the squadron so I'd have more free time to go sight-seeing while ashore.

I had dinner with the guys from the *Lexington* at the Officers' Club and went to the Diamond Patch Shop in Thieves Alley at Yokosuka to pick up my Unit Seven patch and tailor-made rescue patch with the number 61 embroidered in the center of the sling. I had ordered the designs the last time we were in port. This patch shop was rumored to be run by Communists who knew ship movements in advance and made patches for scheduled cruises and ports even before the ships left port.

Each unit designs its own identity patch for wearing on flight jackets and to put on a big display board at the squadron when you return to the United States. Ours is round with a flying red carpet trimmed with yellow on a sky-blue background. The admiral's three stars hang from the front of the carpet with the slogan "Red Carpet Service with a Smile" embroidered around the edge.

Just before catching a cab on the way back from the patch shop, I came across a sailor who was a little bit drunk and was challenging a Marine to a fight. The sober Marine was minding his own business near a group of Japanese. In his drunken condition, the sailor had a false bravado.

I thought it wasn't a good idea for the sailor to cause a big scene and decided to intervene.

He and some of his partying buddies gave me a rough time with verbal threats of physical abuse to come. About this time another 10 or 15 sailors silently surrounded me. Now I was doubting my wisdom of playing peacemaker and U.S. imagemaker. I tried quieting these guys down and shortly the Shore Patrol appeared. The crowd of sailors dissipated.

A couple of sailors came up to me afterward and said they'd recognized me as the *Helena* helo pilot and were standing by to assist me should any violence break out. It would have been reassuring to have known that at the time. Now that would have been a real brawl!

Received a copy of the *Chopper*, the squadron newspaper, today. They had a nice tongue-in-cheek comment about the *Hoi Wong* rescues in the gossip column:

"Flash..!! **Hold the cotton pickin' phone a minute, skipper…a couple of ingenious nephews out in WESPAC, having apparently discovered the unlimited potentials of the "volume business" policy, are insisting on reporting some one-hundred (100) odd rescues. According to substantiating messages originated by COMSEVENTHFLT, and CINC-PACFLT, the number of rescues seem to total 106 or 116."**

"Sir; the Honorable Uncle is requesting permission to notify the Ream Field Bake Shoppe to fabricate five (5) appropriately sized, shaped and decorated cakes for the wing-dingiest cake-eating jamboree in all HU-1 history…!

(Tentative date for cake-eating has been set for 24 October—immediately following personnel inspection.)

Well…what can Uncle Pete say in regards to the incident that hasn't already been said? One little blurb to Lt. j.g. McKinnon….Buster (mach-buster,—that is), it gives this mythical relative great pleasure to be able to say—"I knew him when—"…but if you think you're gonna get a super-sonic ride in an F-100 for each rescue you perform it might be wise for you to re-read the Career Incentive literature. Now, if you can't set-

tle down and behave yourself—stop stealing your fellow squadron mate's rescuees,—and trying to lessen the prestige of the rescue business with production methods, Uncle may have to see if you can't be returned to CONUS and installed in your former position of "writing" PIO material rather than "making" it. What are you trying to do— swamp the system...? (What do you intend to do with all your rescue patch awards...sew 'em together and make a quilt out of 'em...?) CONGRATULATIONS—Units 6 and 7, you are a PIO man's dream.

Tuesday, October 28

Today I spent more money than I think I ever have at one time in my life. My helo blades had some cracks in them so I had to buy a new set of blades for the helo, both fore and aft at a cost of $8000. Fortunately the supply system had them in stock. The old set were packed in special containers and shipped off for repair.

When the new ones are installed, they will have to be adjusted and tracked to ensure that there's the least vibration possible.

CHAPTER FIFTEEN

R and R

Thursday, October 30

TOKYO! WOW, what a city!

John Loomis, my HU-1 buddy from the *Lex*, and I came here yesterday by train. It took less than an hour from Yokosuka. We arrived at the famous Tokyo station—the largest train terminal in the Orient—52 acres. It was constructed in 1914 and was patterned after the Amsterdam station in Holland with Renaissance-style architecture. It evidently escaped the intense bombings of World War II. Each day 1790 trains arrive carrying 490,000 passengers.

We took a tremendous room at the fabulous Oriental Imperial Hotel—the best in all of Tokyo. It had been designed by Frank Lloyd Wright, a world-famous U.S. architect. We were surrounded with ornate fixtures, beautiful carpets, and carved wood walls. The service was outstanding. The cost was $12 a day for both of us. The same accommodation in the States would have cost at least $25 a day.

Strolled around looking at people and gigantic department stores on the Ginza. It's the Rue de la Paix or Broadway of Japan lined with fancy specialty shops, department stores, theaters, restaurants, tearooms, and thousands of odd-looking people.

Today we took a ride in a rickshaw propelled by a bicycle, rather than hand-drawn, on a sight-seeing tour around the city. We saw the Imperial Palace, dozens of Buddhist shrines and statues, and schools with children neatly dressed in blue and white uniforms and carrying backpacks for their books. Everything is so interesting. It's such a different culture from ours.

This afternoon we attended the Nichigeki Theater featuring theme plays, and tonight we were dazzled at the Kokusia Theater, which features lavish Las Vegas– or Broadway-type shows with large colorful costumes and continuous dancing and singing. It was great. Then we walked through the geisha girl section to see what it was like. Our expectations caused a disappointment.

Room service at the Imperial Hotel featured delicious food. We've had some concerns about eating at some of the Japanese restaurants because they look like such little hole-in-the wall joints and don't appear to have the same standards of cleanliness that we have in the States. Besides, we're sort of enjoying living like high rollers at cheap prices for fancy food at the hotel. We even splurged and called our wives in the States. My call for 3 minutes was 5760 yen ($16). John's wife is pregnant and expecting a young one shortly after he gets back from cruise, so they talked even longer.

Friday, October 31

After 3 enjoyable days of R&R and paling around seeing the sights touring Tokyo, John and I came back to Oppama. The whole trip set me back only $50, including two phone calls to the States.

John and I just had a great time together. It's become a good friendship since our first day at survival school. It's funny how a bonding chemistry can develop between people where you just seem to automatically enjoy each other's company and friendship.

The *Helena* is scheduled to leave Monday, although we may be delayed because of a typhoon passing through the area. The weather is brisk at 55°F with lots of rain. However, we had good weather for our Tokyo excursion.

UP-18 is coming together and should be back in flying condition for our deployment on Monday. Tomorrow I'm scheduled to fly to Atsugi. It's an old Japanese airfield used to train kamikaze pilots during World War II. The Navy is using it for the carrier jet operations when the ships come into port. It'll feel good to get in the air again. I haven't flown since last Sunday.

Saturday, November 1

Two U.S. Marine jets collided in a midair last night. Both pilots ejected. I was called out early this morning to fly food and supplies to the rescue and salvage workers at the crash site.

The place was only 75 miles away, but was difficult to locate, and it took 2 hours to get there. Stops at several airforce bases on the way for fuel and directions were necessary. Because of the confusion in pinpointing the crash site, my flight plan got mixed up. I may receive a flight violation unless I can talk my way out of it tomorrow. This goodwill mission may not bring me much good will.

Monday, November 3

The typhoon skirted us, and we left Yokosuka on schedule. Once again I flew out and joined the ship in Tokyo Bay. This undertaking is always a little scary because if I had to down the helo for any reason, the ship would have to wait in the harbor until I got an alternate helicopter, repair mine, or else go off and leave me. There's no room for error.

This time I've got my shiny original helo, UP-18. It sure looks and flies better than the one I had to use the last time we were out.

The *Helena* is scheduled to return Stateside February 11th. I received some bad news/good news today. About the time the *Helena* is scheduled to return to the States, there's a sensitive, top-secret cruise taking place off the coast of a country northeast of Japan. The admiral's staff told me I've been selected to be the helicopter pilot assigned to the ship involved in this special project, so I won't be going back to the States with the *Helena*.

This secret exercise will last 3 to 4 weeks, and we will operate in extremely cold weather. I hate wearing my poopy suit. Heck, I hate cold weather altogether. It delays my return to the States, but assures that I'll be out here long enough that there won't be time to recycle me for another long cruise before my active-duty obligation ends.

CHAPTER SIXTEEN

Testing the Crew

Thursday, November 6

ONE EXCITING THING about life is that you never know from one day to the next what will happen. I found this out yesterday.

I was ordered to fly the helo to the Naval Air Facility Naha, Okinawa, on a routine flight and lay over there several days while the ship performed maneuvers off the coast of Okinawa. The ship's action included a launch of the Regulus missile. They didn't want the helicopter to clutter the flight deck or get in the way of the missile launch or gun firing practice.

While on Okinawa my plan was to take advantage of the time to do some sight-seeing by helo around this historic World War II island. I could also help my crewmen earn their flight skins with additional flight time.

Upon arriving at Naha, I gassed up, loaded the crewmen on board, and was ready to take my first tour flight. I flipped up the start toggle switch. Nothing happened. The engine wouldn't start. It wouldn't even turn over. It wasn't a dead battery because the electrical system was working.

I pressed the starter switch again and heard a small whine. I thought the starter had failed. The crew checked the engine compartment and found that the starter motor itself worked just fine. The problem was some sheared gears that the starter motor thrust deep inside the main engine.

The only way to fix the problem was to replace the whole engine. It was impossible to change parts.

A quick check informed us there were no spare engines at the airfield, nor anywhere on Okinawa, nor on any of the carriers in the area.

I contacted the folks at *Det One* in Oppama. They promised to get one shipped to me.

There's no telling how long it will take to arrive—probably several days. It'll have to be air-freighted. Meanwhile I'm stranded. Then it'll take a couple days to change the engine and a couple of more days to test it and break it in with slow-time flight. My best guess is I'll be stuck here 8 to 9 days.

The ship is going to go bananas when they get the news.

Meanwhile I checked out the accommodations on the base here at Naha. There is a tremendous BOQ with private rooms, semiprivate showers and bathrooms, and refrigerators and hotplates in each room. So this afternoon I went to the commissary and acquired canned pork and beans, Vienna sausages, soup, peanut butter and jam, apple cider, almonds and other nuts, and cookies—all of my favorite foods. I'll have to do my own cooking, but I'll save money by doing it in the room.

It's nice to be off the cruiser, but I hadn't planned to be stranded with these conditions.

The hope is to get the plane fixed in time to ride the ship back to Yokosuka after they complete operations around Okinawa. If I miss the ship, I'll have to wait around until a carrier passes by and I can fly out and hitch a ride back to Japan or to the *Helena*.

One of my old buddies from the University of Missouri, John Bentley, who went through flight training with our Missouri gang, is now with the FASRON 118 squadron. His squadron has an SNB, and he's going to get me checked out in it tomorrow. At least I'll get in a little fixed-wing time. He volunteered to have his men help us change the engine when it arrives.

My proposed sight-seeing of the island of Okinawa didn't happen since we don't have a helo. I feel a responsibility to hang around operations and check every arriving cargo flight to see if my engine was on board. There's only one or two such cargo flights a day.

As we flew into Okinawa from the ship, we passed by the famous suicide cliffs, where an entire division of Japanese generals, officers, and men and their families committed suicide by marching over the cliffs to their communal deaths when they were threatened by the victorious Americans during World War II. I can't imagine that method of thinking.

The much beloved world War II correspondent, Ernie Pyle, was killed here. He was buried in Hawaii, however.

Friday, November 7

No engine yet.

One is supposed to be on its way here from Japan according to *Det One*. It's sure frustrating waiting and not knowing when the engine will arrive.

We figured that at best it will take 2 days to install it and an additional 2 days to test the engine and run off the slow time. After installation of any engine on HUPs, it's required to put 10 hours of flight time on the engine circling an airfield at 500 feet to ensure it's broken in. The purpose is to be sure it works right, and if an engine fails, we have enough altitude to make an autorotation safely onto the field.

Generally new engines fail most frequently within the first 10 hours. A light load is important so that there is less strain on the engine during that timeframe.

The Air Force has one of their flying banana H-21 helos out here, and I've been trying to get a flight on it, but it doesn't look very promising.

To kill time this afternoon, I played a round of golf. Well, almost a round of golf. The golf course had a lot of sand and a little bit of dried grass. My round was lousy. I lost three balls and finally gave up. I didn't have any more golf balls.

Saturday, November 8

Our engine arrived about noon today. That was mighty fast service. The Ops people tell me it usually takes at least a week to get an engine shipped here. We got ours in three days—another advantage of being with the commander of the Seventh Fleet.

The crew began work on it right away. The first project was to get it out of its metal shipping container and deep preservative. The engine was totally covered with cosmoline, which is heavy, sticky grease. It keeps the engine from corroding and rusting while it's waiting to be installed in an aircraft.

It took a long time to clean the engine. Then the crew had to build it up before it was installed in the helo by changing the starter motor, coils, carburetor, and other accessories that are attached to the broken engine.

Sunday, November 9

The crew worked late into Saturday night and all day today. They're going to have that helo in absolute mint condition. It seems to me that the installation is going a little slow. They're doing a good job, but my crew leader works rather slowly, but his work's always been of excellent quality. This will be the first good opportunity for me to judge the type of work my crew does under adverse conditions. Until now, the crew has had very little trouble with the plane, so I didn't have a good idea of their talents. All I knew is they've done a good job of preventive maintenance.

The ship is scheduled to leave the Okinawa area Thursday morning. It'll be a close race for us to get all the testing done in time to get aboard the ship. If the ship leaves without us, I don't know when we'll join up with them.

I arrived early for church service today, so I attended Sunday school as well. It proved to be a good class. Actually the Sunday school was better than the service. We discussed the Beatitudes.

It's a little tough on the crew installing the engine. They're missing some special tools required which are available at the squadron.

I've been hanging around the aircraft a lot while the crew is working on it. But I'm beginning to believe that looking over their shoulders all the time bugs them too much. It's better to let them do their job, leave them alone until they get it done, and then do mine—which is fly the helo.

So this afternoon I visited the Okinawa Times newspaper plant to see how it compared to what we have in the United States. Strangest sight was there are no typewriters in this newspaper office. Since there are 2000 Japanese symbols, all stories are handwritten.

We received word that there was a real fiasco on the *Helena* yesterday. They fired one of their 500-mile range surface-to-surface Regulus I guided missiles to a predetermined target. This is a missile that is normally scheduled to carry nuclear warheads. We have several of the missiles on board. Obviously the nukes to go with them are on board as well. They're located in the hangar at the stern of the ship under the flight deck and are guarded constantly by the Marine detachment. No person on the ship gets anywhere near the missiles in the nuclear storage area unless they have clearance.

Anyway, the way we heard the story, the missile fired properly, shot off the launching rail, and flew subsonically on a predetermined course toward the target. As it left view of the ship it was escorted by a Navy jet fighter that radio controlled the missile to the target. The Regulus is a giant precision-guided cruise missile. If the missile failed to operate properly, the chase plane could shoot it down and prevent it from crashing into a populated area.

Well, the chase pilot thought the missile was veering off course and destroyed it before it got to the target. The telemetry data from the *Helena* said it was working just perfectly.

Now there's a controversy developing over the missile shoot and shootdown. The missiles are very expensive, and it's really rare when a ship ever gets to fire one for practice. So the crew on the *Helena* is furious because they'll never know whether they would have hit the target.

A failure never looks good on anyone's record. The messages are flying back and forth between the ship and the squadron as to what happened and why.

I also heard today that if my helo's engine is ready, I'm scheduled to haul President Chiang Kai-Shek on a VIP trip Friday.

Monday, November 10

The installation of the new engine was finished tonight. I cranked it up, and everything seems to work fine—no oil leaks or any other problems. I sat on the tarmac outside the hangar with the engine running and the rotors engaged for an hour and a half to break it in.

Then came another important test. At the end of the hour and a half, I added enough collective to hover for 5 minutes to see if the engine could take the strain. No problem.

The crew is exhausted, but exhilarated that the installation has gone so smoothly. The real test comes tomorrow when the slow-time flights begin.

Tuesday, November 11

Today was a busy day. Flew 7½ hours of the required 10-hour slow-time test period. To keep the weight low, the helo carried only an hour's worth of fuel. I flew for an hour and then landed to pick up more fuel and then another hour of flight time. The up and down flying took the whole day to achieve the 7½ hours. The engine is running well. The crew did a fine job of installing it. On each flight, I took one of the crew along to get their time in. I also included an ex-instructor from Ellyson Field on the flight who taught me to fly helos.

Tomorrow morning I'll fly the remaining 2½ hours. Then we'll check the sump plugs for metal chips, which would indicate a problem and change the oil.

We should fly aboard the ship in the afternoon. They're pulling out at 1600 hours for a month or more before returning to this area, so we're just going to make it if everything works okay tomorrow.

Helos are still a fascination to many out here, and few have actually flown in them. Tomorrow I'm planning to take along the XO of FAS-RON 118, the squadron that's helped us so much with installing the engine. Boring holes in the sky circling the airfield and looking down is not the most exciting flying in the world, but always involves tenseness, because you never know if the engine might fail on one of these test hops.

Today was Veteran's Day. It honors those who gave their lives for freedom during World War I. Since that war ended on the 11th hour of the 11th day of the 11th month, Veteran's Day was celebrated at 1100 hours today.

CHAPTER SEVENTEEN

Chiang Kai-Shek Special

Wednesday, November 12

ALL THE TEST FLIGHTS of the helo were finished and we transported our gear back to the ship. I've flown more than 12 hours in 2 days. It's taken from sunrise to sunset to accomplish because of the number of short hops and engine checks between each flight. There's a lot of paperwork to complete when there's an engine change. Around the squadron, this task is done by somebody else, but on cruise, as the officer in charge and only pilot, I had to do all the paperwork and reports. My hope is that I did all the engine change documents correctly.

Tomorrow I'm scheduled to go to the *Midway* and spend the night. There's lots of preparation for the big airshow being staged aboard the *Midway* for President Chiang Kai-Shek.

Thursday, November 13

The day was spent ferrying Admiral Kivette and most of his staff to the *Midway* for the visit with Chiang Kai-Shek. The *Midway* is acting like a big iron floating hotel.

There is a minute-by-minute schedule of the events for tomorrow, and everyone involved spent several hours at a conference going over who's going to fly who and exactly how the day should go. The president and most of his high-ranking staff are to be heloed to the *Midway*, which is cruising 4 miles off the northern end of Formosa. I was selected by the admiral's staff to fly Chiang Kai-Shek a month ago, but his coming aboard the carrier was classified. I didn't find out until recently, and couldn't tell anyone, even my crew. I was selected to fly him because the requirements to operate aboard a cruiser are more exact than on a carrier, and I was the most experienced VIP helo driver in WESTPAC, I was told.

Friday, November 14

Today was a big day.

President Chiang Kai-Shek of China or Taiwan, depending on your point of view, visited the Seventh Fleet and saw an impressive display of U.S. Navy might.

It was his first visit to the fleet since the heightening of tensions in the Formosa Strait. It was a "thank you" visit for U.S. support during the crisis between Taiwan and the Red Chinese. That situation has now quieted down.

I was fortunate to have a cat-bird seat to watch a lot of the action. The helo was polished until it sparkled. The new engine has been running flawlessly since its installation several days ago. The white VIP seat and seatbelt covers for the forward and side seats were washed and slightly starched. One of the crewmen had painted a special 8½ × 11-inch plaque with the flag of Nationalist China to slip into the slot on the side of the helo below the pilot's window. We normally use a three-star plaque in that slot to identify when the admiral is on board the helo.

Because the flight safety record of the HUP for the past 12 months has been so awful—12 helicopters lost at sea mostly because of engine failures—a Marine HUS helicopter was assigned to follow behind me. No one actually said what to do if we went in the water, but I pretty well understood what my responsibilities were. First out and to be rescued would be the president, next would be my other passenger, U.S. Ambassador to China Everett F. Drumright, and finally Chiang's interpreter.

There were stories that Chiang supposedly understood English, but never used it around Americans. This gave him the added advantage of sorting out U.S. government tactics before it was interpreted for him. He then had more time to respond.

The day started off with my first launch to the Tamsui Golf Course on the outskirts of Taipei at 0830 hours. There I picked up Chiang's son General Chiang Ching-Kuo and other Chinese officials, and flew them on a 10-minute flight out to the *Midway*. His son was husky and taller than the diminutive Chiang. He also didn't have the same aura or charismatic feel about him. He quietly boarded the helo without special attention. Meanwhile a fleet of Navy and Marine helos were shuttling dozens of Chinese and American dignitaries, officials, and press to the carrier.

After my first trip to the carrier I returned to our temporary heliport at the golf course.

Chief Warrant Officer Mesler from the *Helena* was running the operation at the golf course. When the admiral and his staff had confidence in someone, they stuck with their team and assigned them to all the special duties. A control tower was established to handle all the air traffic between the golf course and the carrier. Mechanics were also imported in case of problems.

At this point, I shut down the helo right on schedule at 0910 hours and waited for President Chiang Kai-Shek to arrive at 0950.

We were to arrive at the carrier at 1000.

The ambassador was already there, and we talked for about a half an hour waiting for the president to arrive.

Then Chiang drove up in big, long, black, shiny Cadillac limousine. I was introduced to the president as the pilot who made the rescues at the Paracel Islands. He grunted several times and smiled. I put a Mae West over his head and cinched the straps. After that I showed him by hand motions the tabs to pull to inflate it. My thought was if he needed it, it would be up to me to inflate the 72-year-old leader's life jacket. He was a short, thin, frail, and fragile-looking man. If we had to ditch, I knew it would take a heck of a lot of effort to get him out of the helo for a rescue.

I seated Chiang in the front passenger seat, and one of the helo crewmen working at the heliport got him strapped in. The ambassador sat in the crewman's seat immediately behind me, and the interpreter was next to him near the rear side door.

I didn't want to frighten the president, or have him remove his fancy, tan-colored hat with all the gold braid on the bill, so we didn't put helmets on him or any of the passengers. Consequently there was no communication through the intercom for the short flight. The helmet obviously would have messed up his neat appearance and hair.

With everyone strapped in, I started the engine, engaged the rotors and did my mag check carefully. Next I double-checked all the engine transmission temperatures and pressures. All looked normal.

I lifted off gently and made a smooth—one of my smoothest—transitions to forward flight. I climbed to 500 feet and then went to 600 feet to give us a 100-foot cushion above "dead-man's curve" altitude in case I had to autorotate.

We headed for the carrier with two Marine rescue helos in trail formation. There were several journalists in the second of the two helos. Cruising at 60 knots, I arrived at the carrier one minute ahead of schedule. Sailors, in their starched dressed whites, stood about 3 feet apart "manning the rail" along the rim or outer edge of the entire carrier flight deck. It looked magnificent. The flags on the mast showed strong winds right down the deck. The honor guards from the ship's company were arranged in a formation that required me to land just forward of the bridge, crossways in the center of the flight deck. The nose of the helo would point to port side, and the helo exit door would face aft, right into the corridor formed by the sideboys.

"School Boy, Gladiator Angel for landing," I radioed.

"Roger, Angel, cleared to land. Winds 270 at 10."

"Roger."

"That's strange," I thought, "stretched-out flags on the yardarms say the strong relative wind is coming right down the deck and Pri-fly says it's coming from the port beam. Must be a local wind down low," I thought, and immediately concentrated on my approach to make a smooth landing.

As I neared the flight deck, the helo started bobbling all over the place. It would take all my effort to control it and plant it on the flight deck. I looked out the canopy, and then it dawned on me. I was looking at the full face of the flag held by the carrier crewman signaling me where to land. The flag was in a stiff cross-wind. The sideboys were in place so their heads were just outside the arc created by the rotor blades. I would miss them if I landed on the spot painted on the flight deck for this occasion.

Here I was landing in a severe cross-wind with the president of China, thousands of sailors, a dozen admirals, and who knows how many photographers and press watching. All the time I'm believing Pri-fly's message that it was a normal-into-the wind landing. If I missed the spot, the sideboys would have been in serious trouble.

It wasn't exactly like a feather dropping, but it was a good and safe landing.

After getting on deck, I breathed a sigh of relief and hit the "disengage" switch. Then I reached for the rotor brake over my head and with my left hand pulled it down to slow the rotors so there would be less exposure to the wind slapping or slamming the blades into the fuselage. It worked out fine. But I was furious with the ship. They lied to me about the winds, and that could have caused an accident.

Chiang crawled out the side door, and the full honors started with a 21-gun salute. Whistles blew, there were salutes, and there were handshakes and smiles from everyone. Then the band played the Chinese national anthem. Admiral Kivette presided, but Admiral Hopwood, Commander of all Naval Forces in the Pacific, and Admiral Smooth from the Taiwan Defense Command were there, too. I stayed seated in the helo.

The crew wanted to know if there was any squawks on the helo. There were none. Chiang started an inspection of the Marine detachment, who were all standing at ramrod attention. The helo was pushed to the forward elevator and lowered to the hangar deck with blades spread. There it would stay until the tour, lunch, and planned airshow were completed and it was time to return Chiang back to his island.

I called Pri-fly on the phone and complained to the air boss that they lied to me about the winds and could have caused a disaster.

"We knew you could handle it, McKinnon, and we needed the helo positioned for the honors. It would have been too much trouble to move the personnel around on the flight deck. Now leave us alone; we've got an airshow to run."

"Yes, Sir, but I'm gonna need better winds to leave here this afternoon, or I won't be able to crank up and leave on time, Sir," I threatened. "You'll have them."

Chiang was escorted to the bridge when the honors had been completed. Then the ship launched 40 planes in 20 minutes—that was fast. Many carried the new deadly heat-seeking air-to-air sidewinder missiles.

Aircraft from the *Lexington* and the USS *Ticonderoga* (CV-14) plus a shore-based Marine unit flew by in a salute to the Chinese president.

This was followed by demonstrations of low-altitude bombings, missile attacks, strafing, and napalm firebombing by both the jets and the propeller-driven ADs. Wave after wave of jets scored direct hits on simulated targets in the mile gap of water between the *Midway* and the *Helena*, as they cruised in formation.

It was an awesome display of air power. I watched most of the show from the hangar deck. When it was over and all the *Midway* fighters safely recovered, we positioned the helo on the flight deck ready for departure. Chiang ate lunch and inspected the troops again. Sailors in their white uniforms formed a corridor to the door of the helo.

We were scheduled for a 1430 (2:30 P.M.) departure. I was strapped in the helo by 1400 and waited while the departing ceremonies took place. We played it safe by using auxiliary power from the carrier to start the helo. If the battery were dead, we could get it started and still fly without delay.

While the crew was strapping in Chiang, the American ambassador, and the interpreter, I cranked up the engine and was checking the controls on the helo. I couldn't get my hydraulic boost for the controls to work.

I flipped the switch back and forth three or four times. It's impossible to fly this bird without the boost except in emergency conditions. It takes both hands to push and pull the cyclic without the boost.

Things became tense.

The pomp and circumstance to get the president in the helo had reached a crescendo to the final step of departure. The engine was running, the crews had strapped him in, the excitement of departure was building, and I was stuck. My pulse rate increased. What a time for the helo to be uncooperative.

Then it dawned on me. We used auxiliary power to start the engine. When you do that, you don't turn the battery switch on until the power plug is pulled out of the helo. The boost doesn't work with the battery switch off. I reached over to the pedestal and flipped the battery switch on. The boost came on and the controls immediately responded. Boy, did I breathe a big sigh of relief.

The rest of the preparations were anticlimactic as the crew finished their preflight chores.

"School Boy, Gladiator Angel ready for launch."

"You're clear for launch, Angel. Winds calm."

And they were. The ship was steaming on the correct course.

I lifted off and circled the ship at 200 feet so Chiang could see the white puffs of the 21-gun salute they were giving him and then we headed for the beach.

The SAR (search-and-rescue) helo was tucked in right alongside us. We arrived at the helo pad. The flight was uneventful.

As I reached over and undid Chiang's seatbelt latch, he reached out so we could shake hands. The other passengers leaned forward and shook hands with me as well.

We sat there disengaged with the engine stopped. There would be no accidents during the unloading. As Chiang walked aft, I grabbed my movie camera and leaned out the window to photograph his departure from the helo. Chiang saw what I was doing, walked around to my window about 10 feet away, stood there for a moment smiling, saluted me, and waved. I continued to film. Then he slowly turned and walked to his limo. I probably got better pictures than anyone on the ship. I only hope they turn out okay.

There was an army of guards and secret service at the helo pad watching out for his safety. They were hidden everywhere. The story was they knew nine ways to kill a man—I didn't doubt it—but the only way I saw was the machine guns they carried. I'm sure everyone was relieved as I was that the journey was over.

After Chiang left, I made several shuttle trips between the *Midway* and the heliport carrying various VIPs, including Chiang Kai-Shek's son and other generals.

I finished the day by flying all our gear from the *Midway* back to the *Helena* and had a great night's sleep.

CHAPTER EIGHTEEN

Goodbye, John

Saturday, November 15

TODAY I LOST my enthusiasm for flying. My best friend in the squadron, John Loomis, was killed flying in his helicopter.

He was flying 53-year-old Rear Admiral Leonard B. Southerland, commander of all the aircraft carriers here in the western Pacific, and a member of the admiral's staff, Commander John Coulthand, 42. They were flying about 800 feet over Okinawa on a routine trip from the *Lex* to Naha, the auxiliary airfield on Okinawa, where we changed our engine just 2 weeks ago.

They were flying along in clear weather when the rotor blades snapped off and scattered in different directions. Witnesses said the rotorless helo plummeted to the earth near the Machinato Quarter Master Depot and burst into flames at impact. All three were killed instantly.

John was one of the best pilots in the squadron and certainly the best aboard the *Lex*. Of all the guys in the *Lex* detachment, the best of them dies through no fault of his own. Where's the fairness?

I've tried to figure out what his final thoughts might have been as he futilely tried to manipulate the controls the last 20 seconds of his life. I doubt if he was aware of exactly what happened.

We sure had a great time together in Tokyo. Death was the last thing on our minds. That trip was so full of life. Who would have ever dreamed that just a few days after we parted company John would be dead? He was only 23. I'll miss him and the fact we won't be able to share adventures in the future like we did in the past.

The ship has been sensitive enough to cancel all flight ops. I'm not sure whether it was out of concern for safety of the passengers I'd be flying, or in sympathy for me. Either way, I just want to be alone for now.

The helicopter can give life, but it also can take it away.

Now interservice rivalry has developed over the Chiang Kai-Shek visit. The Navy Public Relations sure dropped the ball. It was reported in the news coming back to the ship that Marine helos carried President Chiang Kai-Shek to the *Midway*. Reality was that they were temporarily based on the *Midway* to ferry newspeople. It was a Navy show all the way with Navy helicopters carrying all the VIPs. The PR (public relations) battle never stops. I'm sure somebody is getting reamed out for the lousy press coverage the Navy received and some Marine somewhere is bragging about his great PR coup.

We've now set course for Japan. We're planning on a stop in the city of Iwakuni in southern Japan. We're tentatively scheduled to spend the Christmas holidays in Yokosuka, if there are no fireworks down south.

Sunday, November 16

There's no word on what caused John Loomis's crash.

No flying was scheduled again today, and I'm glad.

My crew and I are carefully examining my plane for any defects whatsoever to try to avoid anything happening to us, but it's tough when you don't know the problems you're looking for. Hopefully we'll hear something soon so that my crewmen will know what to look for on our bird.

On top of it, I feel lousy because of the flu shots I got a couple of days ago. Most of the men on the ship feel the same way. The dosage may have been too strong. Maybe the cure is worse than the disease.

The admiral had a Marine Corps general on board the flagship today. I ran into them on a ladder.

As I saluted, Admiral Kivette in a lighthearted way told the general, "Here's my helo pilot, he's got a couple of loose rotor blades up there, but he's alright."

He was trying to be humorous and to lighten the load I was carrying for John, but I still feel heartbroken to have lost my best friend and not even know the cause.

It's sunny, but cold and crisp today. Winter is coming on. We're passing through the Inland Sea of Japan en route to Iwakuni, where we expect to anchor late this afternoon.

The ship is navigating through the straits, which are about 500 yards wide—a tricky feat for a ship this large. For a change, the sea is nice and smooth.

Monday, November 17

I flew off the ship this afternoon to Iwakuni Marine Corps Air Station here in southern Japan. Some minor repairs were done to the helo, and I looked around the air station. The admiral is a history buff, so I've received special permission from the Japanese government to fly him tomorrow morning over Hiroshima to see where the first atomic bomb was dropped. It's a sacred shrine, and normal flying over the site is prohibited.

Tuesday, November 18

At sunrise the admiral and I met at the helo for our departure. My back has been out of joint for 2 days, and I can't even bend over to tie my shoes. This caused me to get a little careless. I was distracted enough to not complete preflight in my usual manner, especially when I'm in a strange place without my crew. I climbed in with my painful back and tried to strap in.

Then one of the air station flightline guys reached under the helo and handed me the pitot tube cover. In my preflight of the helo I'd forgotten to take it off. I was mad at myself. Normally my crew does all that for me, and I've become a little spoiled. This was my reminder to be more careful. Flying without an airspeed indicator would have been all by feel and a little difficult.

With a feeling of solemn interest and excitement we took off for Hiroshima. The ship left port at 0600 hours and proceeded up the Inland Sea about the time we were ready to fly. The city of Hiroshima has been rebuilt and doesn't look like it was leveled by the "A" (atomic) bomb.

I wasn't quite sure where we were so I kept a sharp lookout for the city's Peace Memorial Park, location of the symbolic A-bomb dome, a ruin from the atomic attack.

When we flew over it, it was about the same time of day the bomb exploded at 1800 feet above the ground on August 6, 1945—0815 hours. An estimated 140,000 people died in the blast, firestorm, and radiation from the attack on this western Japanese city that was a major military center during World War II. Much of the city appeared rebuilt as we flew over it.

Three days after the Hiroshima attack, the United States dropped another atomic bomb on the coastal city of Nagasaki, not far away, which killed 70,000 more people. The admiral and I didn't get over there.

It was that second bomb dropped 6 days later on August 15, 1945, that caused the Japanese to surrender unconditionally. That was 13 years ago. These data made me reflect on the impact of all-out war as I changed course to head northwest toward the flagship.

Next we flew over Kure, the largest shipbuilding center in Japan. Being the admiral's pilot provides lots of interesting experiences. I let the admiral fly a good portion of the time, including trying to hover over the landing circle on the fantail of the ship. He gets a kick out of it, and I really enjoy teaching him. He did a good job for a beginner.

I turned off the electric boost on the controls. This requires the use of two hands to fly. The admiral pushed, pulled, grunted, and groaned, and after about a minute he turned red in the face and asked for the boost back on. The whole thing was so funny, I didn't stop laughing for about 5 minutes. Fortunately he's got a good sense of humor. We're getting along great.

This afternoon I'm to take him to Itami Air Base, where he'll catch a plane for Yokosuka. The ship arrives in Yokosuka tomorrow afternoon. Itami is home of the Shin Meiwa Company, an overhaul-and-repair facility. All the HUPs in this part of the world are overhauled in that facility. The only other overhaul facilities are in San Diego and Jacksonville, Florida.

After letting the admiral off, I'm to rendezvous with the ship as it passes Kobe—the city famous for tender Japanese beef.

During a quiet time in my stateroom tonight, I wrote to Nancy Loomis, John's expectant widow. Didn't know exactly what to say, but felt it was important to say something.

Tuesday, November 18, 1958

Dear Nancy:

It is with a heavy heart I write to you to say that I also was shocked to hear of John's death. John was my best friend in the squadron and we had shared many enjoyable and exciting experiences in the year we were in HU-1 together. As a matter of fact, just a couple of weeks ago the two of us toured Tokyo together where we both called our wives one night from our hotel room.

I know you miss him very much indeed, Nancy, and talking with him I know he loved you dearly. You certainly were number one in his life.

John loved to talk about being an expectant father and as I see it this is one fortunate circumstance that God gave you, in that you will have a part of his personality and characteristics around in the person of your new child.

Once again I offer you my deepest sympathies in your time of grief. May you have the faith in God and receive the strength He gives which you need so much at a time such as this. If there's anything I may do for you while here in the Far East or later, or if my wife in San Diego may be of any assistance, we'd feel honored if you'd call us.

Thursday, November 20

We're back at Oppama today. I left the ship yesterday before it docked, and during the short flight, my boost system went out. The mechanics are still having trouble getting it fixed today. It made for some tough flying.

The crews from the *Midway* and the *Bennington* are here, too.

The accident investigators discovered the exact trouble that killed Loomis. A bearing in the aft rotor head burned and melted because of the lack of lubrication. It dephased the rotors, causing seizure of the aft rotors. In essence what happened is that the drive shaft to the rotors suddenly froze in midair, which snapped all the whirling rotor blades off the helo at 800 feet altitude. Then it just plummeted to the ground like a rock where it turned into a fireball. There is an aircraft service change kit designed to rectify the situation by installing a greasefitting to be lubricated daily. It just hadn't been installed in that particular helo yet.

Now all HUPs are grounded until they do have it installed. My plane had it put in the last time we were in Oppama a couple of weeks ago.

Until now there has been a deep-seated concern over what caused the accident and whether it would happen to me. When a pilot is flying and knows somebody's rotor blades in an identical aircraft snapped off and he doesn't know the reason, the pilot is mightily concerned. Now a lot of my fears have been relieved, and I'm enjoying flying again.

The top-secret part of the cruise that I was scheduled for following this cruise has been canceled. Hooray, I'll get to go Stateside 4 weeks sooner.

Friday, November 21

I went to the specialty shops in Thieves Alley in Yokosuka and ordered a baseball cap for the admiral with Honorary Helicopter Pilot of HU-1 written on it and lots of gold braid on the bill.

My plane is still down for repairs. The weather is rainy. So it was an easy day hanging around the hangar.

Saturday, November 22

I went by the hangar and tested UP-18, and the boost system seems to be working. It's certainly great to have a headquarters out here with good mechanics to fix our helo while we're in port.

The guys on the *Lex* are furious with me. Seems the quick-delivery service from Okinawa for our engine was a little too quick, even for the admiral. The engine installed on my helo was destined for the *Lex* helo detachment, who also were in desperate need of an engine. They'd been waiting for 2 weeks. We had stolen their engine. We didn't pay much attention to shipping labels, we just assumed any HUP engine was meant for us. In this case possession is 100 percent ownership. There'll probably be a paper trail to untangle since our engine is now floating around somewhere.

I've got 20 hours left to fly on this bird before it's due for a major check. A "major check" means we've got to completely disassemble and inspect the aircraft and all the major components to find any cracks, faulty fittings, or leaks.

I thought we were going to do it now in Oppama, but we'll have to wait until our next visit in about 3 weeks. We don't have time to fly 20 hours and take the 4 days necessary to conduct the check prior to the ship's departure.

Monday, November 24

I checked over the enlisted evaluation sheets of my crew for their semi-annual marks on behavior performance, and knowledge. I think I was a little too liberal with their marks on the rough sheets, and so I've lowered them a notch here and there on the smooth copy that goes into their permanent record. They did a great job on the engine change and generally keep the helicopter in good repair and condition; nevertheless I think a 4.0 is a little too high for everybody.

CHAPTER NINETEEN

Typical Operations

Tuesday, November 25

BAD NEWS. ALL HUPS have been grounded again. Here at *Det One* a HUP sheared a pin on a sprocket that holds the longitudinal control cable. Fortunately it was on the ground just before lifting off when the incident happened.

In simple language, that means a sudden loss of fore/aft movement of the controls while in flight could result in an uncontrolled loop or noseover. The consequence of such activities would be fatal.

So once again all HUPs worldwide are being inspected for this new malfunction. It doesn't bother me much, not like my concern of John Loomis's problem. Knowing the problem is one thing. Fear of the unknown is another.

This will put a crimp in my schedule. When the ship pulls out, it is headed for a visit to Hong Kong. This is my only chance to get there. If we can't have the helo ready, I'm trying to work it out so I can ride the ship to Hong Kong without a helo, and then fly back to Japan from there to test the fix.

Piasecki Helicopter Corporation, the outfit who manufactured the HUP, has some experts over here discussing various difficulties we're having with the helos. I spent the day around the hangars listening to the discussions with the maintenance guys as they try to figure out what to do.

Wednesday, November 26

At the hangar again tonight—the *Det One* mechanics have been working trying to get a helo ready for the *Midway*. All the checks and special safety inspections demand a lot of work and effort. The carrier naturally has first priority on the first helo to be fixed. The carrier is ready to pull out, but with no planeguard helos there'll be no flight ops. The *Det One* mechanics literally swarmed over the helos and are working feverishly to get them in flyable condition.

The carriers use destroyers for planeguards at night when the helos can't fly, but insist on helos for daytime ops. There's a great deal of pressure on the mechanics who are virtually working around the clock. No one gives much thought to the helos on the boat—until they don't have one.

I gingerly flight-tested one of the *Midway* helos and it seems to work okay. I get the third one put in flyable condition and it's going to be touch and go if it can be put back together in time for the ship's departure tomorrow afternoon.

Thanksgiving, November 27

I attended church service on the base this morning. We Americans have so much to be thankful for, and being here in a foreign country helps you place real value on our freedoms—freedom of thought, speech, religion, actions, and worship.

I just learned that the most inspirational Christian chaplain I ever met in the Navy is getting transferred from Oppama. I need to share with him how great I think he was at the Thanksgiving service.

Spent all afternoon today at the hangar working on the helo. We got the third one up just before dark. We're on such a tight schedule that I used the test flight to ferry it over to the cruiser on the assumption nothing would go wrong.

What a hairy landing! There are two carriers moored on opposite sides of the large carrier wharf and another cruiser plus the *Helena* docked at a smaller pier between the carriers. As we inched along, the mooring lines crisscrossed to the stanchions on the pier. I noticed the toll of rust the heavy seas were inflicting on the carriers' sides near the waterline where heavy-duty bumpers protected the giant ships from rubbing the dock.

I just barely had room to fly over the water between the *Helena* and one of the carriers, then turn 90 degrees and nose onto the cruiser flight deck. It wouldn't have surprised me to have clipped the tip of the rotor blades against the carrier.

There was no choice; I had to be aboard the ship before dark since the flagship was pulling out at midnight.

Paul Senden, an old Delta Upsilon fraternity brother from Missouri, stopped over from the carrier USS *Yorktown* (CV-10), docked alongside us. He's in VS-37 flying the *Stoof* (S2F) until May.

We sat around and reminisced about our college days, and discussed the flying we're doing now until his ship was ready to pull out. They went through Hawaii on the way here for their ORI (operations readiness inspection). While there he married a girlfriend from the States, an American Airlines stewardess. She plans to be a "seagull," or wife who follows a ship or squadron to foreign ports. Being a flight attendant gives her pass privileges, so her trips cost nothing.

I had dinner at the Oppama "O Club" with a group of squadronmates, then was back aboard the ship by 2000 hours. I was in the sack when all the blackshoes shoved off and we were out to sea.

I also talked to our guys from *Benny Boat* (*Bennington*) today. They lost a HUP yesterday. Dave McCrackin was flying in planeguard position and had an engine failure. He swung around and pointed the helo in the direction of flight, autorotating into the water without any serious problem. Both he and his crewman climbed out through the side window and rear door without a scratch. They were rescued by a destroyer. The second helo was stored in the hangar deck and couldn't get airborne in time for the rescue.

He's getting good at ditchings. It was just about this time last year that he was flying planeguard off the California coast and had an engine quit. That makes nine HUPs that our squadron alone has lost in the last 12 months. Not a very good record. Only two were caused by pilot error; the rest were material failures.

At $319,000 a plane, it gets expensive. (By comparison, 1993 prices for an H-60 Black Hawk is $5.9 million, H-3 Sea King is $6.4 million, and CH-53 Super Stallion is $19 million.) The helo accident rate is higher than any other type of aircraft. Fortunately our fatality rate isn't as high as the jet jockeys. That's one reason I never let my hands leave the controls. If I have an engine failure, I'd definitely try to get the collective down fast.

National Geographic Magazine is doing a feature story on the Seventh Fleet. All the Red Chinese activity helped create an interesting feature.

The reporter-photographer team arrived by COD on the *Bennington*. I was to fly them to the *Helena*, but had one little problem. My passenger seatbelt wasn't big enough to stretch around reporter Franc Shor's waist. He was a short, affable man, but weighed at least 300 pounds. He said he didn't care about the seatbelt, but I did.

The ship called over the radio and demanded to know when I was going to launch.

The pressure to get launched was mounting.

Finally I jury-rigged one shoulder strap diagonally across his chest and attached it to one lap seatbelt. It held him fairly stable, but I don't know what would have happened if we had had to ditch. His partner, Bill Garrett, the tall lanky photographer, sat in the crew seat behind me. (Each of these two men, Shor first, then Garrett, were later to become editor of the *National Geographic Magazine*.)

The commander of the Seventh Fleet received special permission from the British for me to fly Garrett around Hong Kong harbor to take pictures as the *Helena* arrived. Flying in the area is prohibited because the center of Hong Kong is only 10 miles from Red China, and the U.S. Navy isn't on friendly terms with the Reds. I spent time reviewing my charts and memorizing a few checkpoints to know Hong Kong's boundaries, as I wasn't too keen on getting shot down.

I flew the admiral over and back from the *Bennington* today. As he boarded the helo, I handed him the baseball hat with gold wings embroidered on it, the rim full of gold braid and the words, "Honorary Helicopter Pilot."

Aboard the *Helena*, he wore it as he got out of the helo and showed it off to all the usual contingent of honchos greeting him, including his chief of staff and the ship's executive officer. I could see the XO's eyes roll back as he glared in my direction. I'm sure he said to himself, "That McKinnon, he's gonna try to pull an end run around me by getting the admiral to wear a casual baseball cap on the flagship so he can wear his. But it isn't going to work."

"Right idea, anyway, XO."

Friday, November 28

We're barreling along at 22 knots headed for Hong Kong. Our normal cruise speed is about 15 knots, so with all this speed, this ship is really vibrating.

Monday, December 1

It's a beautiful day with clear skies—great for the photo hop as we enter Hong Kong harbor. I had the helo topped off to 500 pounds of fuel—enough for an easy 2-hour flight for Garrett and I.

We launched about an hour and a half before the ship docked. The ship would be maneuvering in the confines of the harbor and didn't want even the possibility of recovering a helo until a half an hour after docking. If I had a problem, my orders were to make an emergency landing on a finger of land sticking out of the harbor known as Kai Tak Airport.

We had the most marvelous scenic flight around Victoria Island near the harbor and flew from 5 up to 1000 feet. The harbor was full of fabled Chinese junks with red or musty-colored gray, nearly square sails as they sailed in all directions in the harbor. Freighters with dozens of foreign flags on their sterns were docked on the wharfs on both sides of the harbor, and scores more were anchored in the channel area. Small boats and ferries were stirring up wakes as they sped between the lush green hills of Victoria Island and the poorer Kowloon area. Garrett was busy snapping pictures with his set of 35-mm Leica cameras. He concentrated on shots with the *Helena* in the foreground, with views of Hong Kong and the harbor in the background. I don't remember the ratio of pictures taken to pictures used in *National Geographic Magazine*, but it is something like one out of every 2500. He shot enough film for at least two pictures to be used. He was wearing a fancy vest with pockets bulging with film and lenses.

In the helo we were able to frame some great shots with Chinese sailing junks in the foreground and the beauty of Victoria Island in the distance. What an exhilarating flight!

We landed after the ship tied up to a pier. The mail arrived and liberty commenced.

With the mail today, one of my roommates received a "Dear John" letter. His wife is getting an annulment. They got married 2 weeks before we deployed from Long Beach, and now she doesn't want to wait for him any longer. When he first told me what he had done, it seemed like an idiotic thing to do in the first place. One of my other roommates' marriages is on the rocks, too. Cruises are hard on marriages—even more so for the enlisted guys.

This Navy life isn't anything glamorous, it requires hard work by the crews, at least 12 to 14 hours a day, including watches at sea. The sea periods include not only long cruises like this one, but lots of short 1- and 2-week cruises to prepare for the long cruise. There's a lot of away-from-home time, which puts a tremendous burden on the wife. On top of that, wages for the enlisted crewmen barely provide enough money for their families to pay rent and buy food. Aboard the *Helena* they sleep in large bunk rooms with tiers of beds stacked four high with barely enough room for a man to walk between the stacks.

My letter from home today emphasized that John Loomis's crash had a frightening effect on my family. My wife was driving her car and heard a newscast on the radio talking about a San Diego Navy helicopter pilot who died in a crash while flying an admiral in the Far East. The announcer added that the name of the pilot was being withheld until notification of next of kin. She knew that flying an admiral was exactly what her husband was doing, and she hadn't been home to receive any phone calls or visitors.

Devastated, she sped to my parents' home. They turned pale on hearing the news. They sat around the dining room, numb, fearing they had lost a son, but they weren't sure. My wife didn't want to go home. They didn't know what to do.

Finally, my dad, owner of a community newspaper and former Congressman, called a friend at the morning daily, the *San Diego Union*, and asked him to check the wire service stories.

They waited tense moments.

The name had been released. It was Lieutenant jg John Loomis—not Dan McKinnon.

Anxious hours gave way to relief for my family—and sorrow for an unknown family.

Wednesday, December 10

We're on our way to Manila. The ship has been rocking and rolling in the heavy seas. I almost got seasick last night, but managed to eat dinner and felt better. The movement is so intense that it's hard to sleep rolling from one side of the bunk to the other. There's no way to brace yourself. What would be ideal is a fence to hold you in like I used to have in my upper bunkbed as a kid.

Four hours flying this afternoon were spent searching the choppy seas for a man who fell overboard from a tanker in our formation. I wonder how long a guy can survive in rough seas with whitecaps spraying mist without a life jacket. I hated giving up, but I suppose his life expectancy wasn't much more than 15 to 20 minutes. The tanker sent over a complimentary message for my efforts.

Saturday, December 13

We spent 3 days in Manila Bay. The ship staged several receptions and social events for the admiral. Flying to Sangley Point and an overnight trip to Cubi Point got all the flight time in for my crew. I tracked down all my gear and paid gas and other bills I left behind when we left suddenly for the *Hoi Wong* rescue. I even visited with my T-34 instructor from flight training in Pensacola, Lieutenant Goodwin.

A letter from my relief wanting to know about the situation out here at least told me the squadron is planning an end to this cruise some day.

Tuesday, December 16, Keelung

Seventy days to go! The greatest news I've had in a long time is our schedule to return CONUS is firm. The only thing that could change it is war. Things with the Red Chinese have quieted down. The cruise will be 2 weeks longer than we thought, which makes for a 7-month cruise, but at least our return is firm. The schedule's been changed so many times, one has to live from day to day. The rest of the cruise will be more like a flagship's should be. Our schedule is something like this:

> December 19 to January 5—Yokosuka, then a tour around Japanese ports until January 26th. The admiral and his staff transfer to another cruiser, and we leave for the States February 7th and arrive February 23rd.

The scheduled trip to Saigon is canceled because of all the activities in the Formosa Strait. It's supposed to be a pretty city with French influence. Maybe I'll get there some other time.

The flying schedule has slowed down. The high-ranking officers on the admiral's staff are afraid to use the helo, although the admiral is still enthusiastic about flying helicopters.

Our ship's senior officers still refuse to fly in my machine. It would help them understand my operational problems if I could convince them to take an orientation flight. They think helicopters are only a type of experimental aircraft.

I intend to stay on the ship for the 2 days in Taiwan. I've pulled duty all day in case anyone needs the helo. At night the guys exhaust themselves on liberty. I decided not to do that.

Tonight the Chinese military forces presented a variety show at a local theater and wished to host the men from the *Helena*. Seventy seats were reserved. It was a thoughtful gesture, but a couple of us officers were able to round up only 17 guys. All the Chinese singing and dancing was great even though we didn't understand a word. It wasn't anything fancy, but the Chinese sure tried hard to entertain us.

Friday, December 19

The intensity of fleet operations has subsided—no strategy conferences and flight ops from ship to ship.

I've tried again to get checked out as a CIC watch officer, but every time there are real flight ops, I have to bug out. My desire isn't working out.

Despite the fact that I wanted to qualify as a communications watch officer and I had the required top-secret clearance, the communications guys have nixed the idea. They are a different and arrogant breed—they feel compelled not to share anything and act high and mighty as if it were their charter to know more than anyone else.

Their concerns over the fact that if I ever went down over enemy territory I could compromise the ship's way of operating caused them to deny me CIC access. So I'm getting lots of reading done. I just finished a book on Hannibal and his battles and conquests in ancient Rome.

Saturday, December 20

We arrived in Yokosuka today. I flew off the ship over to Oppama at sunrise. We're planning a major check on UP-18, including removing the blades for a thorough inspection and checking over the engine and entire helo to find any problems. I flew the crew around a couple of hours to get their flight time in and to use up all the time needed before the check.

I flew more hours this month than any since I've been in the Navy. My goal is to reach 500 helo hours before leaving the Navy next November. I've got about 370 now.

The guys from the *Midway* are here, so it won't be completely lonesome for Christmas. We'll be lonesome together.

LCDR Bill Hart at *Det One* has been inviting some of us strays hanging around the BOQ to his Japanese-style house for a home-cooked meal. Mrs. Hart had a way of serving the same basic food served aboard ship that made it individually attractive without the mass-produced taste.

Tuesday, December 23

I saw Bill Garrett from *National Geographic* at the Navy Exchange today. He talked me into buying a Nikon SP camera. It cost $200, or about a quarter of a month's pay. He claims it's the best-made 35-mm camera and would cost about $400 in the States. He also gave me a bunch of his reject slides from our flight over Hong Kong harbor. Even though they were his rejects, I thought they were great.

We had a sandwich for lunch. Garrett was always full of interesting stories. We talked about our ship's chaplain's visit to whore houses in Hong Kong.

"I was sitting around about 5:00 one afternoon in the lobby of the famous old Peninsula Hotel over on Kowloon having a drink with some admirals," he related. "They were talking to a gal named Mimi who resembled a classy Suzie Wong and was believed to be a Communist spy.

"About seven that evening Chaplain Kevin Keaney, the *Helena*'s Catholic priest, wandered by in civilian clothes and asked if he could join us," Garrett related.

"We sat around another hour and a half captivated by Mimi's stories of life in this Oriental city.

"Then I told the chaplain, 'Let's go eat,' and we were off for a Russian restaurant," Garrett said.

"'Can I join you?' Mimi asked.

"I thought someone else would take her to dinner.

"We had chicken Kiev and talked about the seedy side of life in Hong Kong. Mimi seemed to be an expert on the subject.

"Then Chaplain Keaney blurted out, 'I've never visited a whorehouse. Would you show me around?' Garrett added.

"And we were off, not to do business, however. Mimi chaperoned us. As we visited the women in the hole-in-the-wall places designated as bars, they cheerfully sold us drinks and we watched how the women operated, except the bedroom action.

"After two or three places, it finally dawned on the chaplain he wouldn't be a very good example giving Mass to the men if any of the *Helena* sailors saw him there, so he said a prayer or two and we did a quick exit. Fortunately he was never spotted. I assumed he gained a better understanding of some of the sin he had to deal with," Garrett concluded.

The *Midway* guys went on a 3- to 4-day trip to the mountains so I'll be the only guy around from HU-1 on Christmas.

Wednesday, December 24

It's lonesome to spend Christmas Eve in a foreign land. It really doesn't seem like Christmas in a land that worships Buddha statues and Shinto shrines.

Bill Garrett came for dinner, and we went to the Oppama Chapel. He was impressed with the Japanese choir singing Christmas hymns and carols. My usual gripe is the audience doesn't get to sing enough hymns. With my voice, besides the shower, the only other place I can sing without any negative comments is in church.

A nice Christmas card from Bill Yoakley, my first rescuee, arrived today.

This morning I read the AAR (aircraft accident report) on John Loomis's accident. He never had a chance.

I tested my helo, but found it still has a couple of things that need fixing. We'll get that done after Christmas Day.

Christmas Day, December 25

I celebrated Christmas alone in my room by slowly opening presents shipped from home. I had never realized how important mail at Christmas can be to a serviceman stranded in a country far away from home or aboard a ship. It was a morale booster to receive all the packages.

The BOQ kitchen is closed, so Bill Hart invited me for an afternoon dinner. I took along some wrapped gifts for his three kids. I took a train and bus to reach his home, which got me to thinking how spoiled we Americans are with our own cars, always depending on them to get us around. The majority of my travel here has been by train, by bus, or on foot.

Monday, December 29

Today I learned a way to save money. There's a ham radio operator at a building adjacent to the Yokosuka heliport. I took a training flight over there, shut down, and had a good chat with the guy. He was impressed that I flew the helo over just to talk with him. His hobby is talking to other ham operators in the States, then patching a phone call to the home of servicemen they're trying to help. The only charge is a long distance call from city to city in the States, and not for an expensive overseas call. He said he'd place a call for me tomorrow. I plan to take the train back to do that.

Thursday, January 1, 1959

New Year's Day is quite a celebration in Japan. The women and children are all decked out in kimonos with the traditional hairstyle, much of it accomplished by renting elaborate wigs.

There's also the tradition of paying one's debts, cleaning the house, and taking a bath to get a fresh start for the coming year. Then there's the traditional feasting on black peas symbolizing health and eating dried fish representing success and herring roe signifying many children.

CHAPTER TWENTY

VIP Operations

Friday, January 2, 1959

IT SNOWED ALL DAY YESTERDAY and last night. Today the front passed through, leaving an inch of snow on the ground. It's beautiful, but not the kind of weather I expected in Japan.

My UP-18 is working fine, and all the bugs have been corrected.

I gave 3 hours of area check-outs to the guys from the USS *Bonhomme Richard* (CV-31) who just arrived in the Far East, plus got in the flight time for my crewmen.

When I arrived at *Det One* the new O-in-C came dashing out to the helo mad as hell and sputtering: "Where've you been? We wanted to use your VIP helo to haul the admiral off of one of the carriers and you're just out flying around."

He continued his exasperation in front of everyone.

Everyone likes my helo because it's like brand new—all cleaned up and polished. Since it's assigned to me and the *Helena*, I consider it my personal property. I apologized and told him I was sorry. I didn't know he wanted to use it. What more did he want?

Well, he didn't know, but something else besides my apology. He restricted me from flying until Monday. I'm not sure he even has authority over me. We're pulling out with the ship Sunday, so it doesn't mean much. But the tone and mood at *Det One* sure has changed with the new O-in-C.

Saturday, January 3

I ferried the plane to the ship today, even though we're not pulling out until tomorrow night. The XO is nervous about a forecast of bad weather tomorrow that could keep me from getting aboard. He doesn't want the admiral disappointed.

I'll stay in the BOQ tonight and bring my personal gear over in the morning. We've been dropping so many helos in the drink recently, I've started transporting my personal stuff by ground transportation so I don't lose it if the helo goes in the water.

Just as I landed, the admiral's Marine Guard, with his 45-caliber sidearm on his hip, came to the flight deck and told me the admiral wanted to see me. I was a little nervous since I was not aware of anything I had done wrong.

I went to his cabin. He explained that later this month we're going on a goodwill tour of key cities in southern Japan. He's putting his schedule together and has some long-distance helo trips planned. He was in a serious mood and wanted to be personally assured it would be no problem for me to get him around.

Many of the trips would be over 75 miles long, mostly over land. I told him there would be no problem. But I thought to myself, I'm glad I brought along all the charts I pilfered from the squadron. I'll need to do some studying since all the navigation will be DR (dead reckoning), and I'd sure hate to get lost or run low on fuel with the admiral aboard.

He kept wishing me "Happy New Year" in his usual, friendly manner. It was a boost to my spirits. It appears we'll get quite a bit of flying this time out.

What a difference one man can make. The whole attitude at *Det One* has changed since the new O-in-C replaced LCDR Fortin, who was transferred to NAS Kingsville. Each shipboard *Det* has learned to adjust and operate under the different circumstances aboard ship. All the helo detachments are assigned and report to the ships we're stationed aboard. *Det One* has basically helped with major repair work, logistics, and assisting wherever needed. The new skipper is trying to require more reporting to him by deployed units. It just isn't the helpful, friendly outfit it was under the old O-in-C.

The new guy isn't liked by any of the units that I've talked with. And no one at *Det One* has the same old "can do" spirit anymore.

Word from the squadron, at the end of the year, is we have a total of 801 rescues in 10½ years.

Sunday, January 4

We are under way again tonight, off for a visit to all the important southern Japanese cities. Tomorrow morning we'll arrive in Nagoya.

I'm upset with the crew. The helo isn't up to the usual neat and clean standards, so I've restricted them to the ship until I'm satisfied with its condition. The idea of no liberty at the upcoming exotic ports has them working hard.

I gave Dunlap a royal chewing out. I'm more convinced than ever that you can't turn people loose and expect top-notch performance. They need continuous accountability.

Forty-four more days and counting until the cruise is over. We're due in the States at 1600 hours, February 16th. We'll return directly to Long Beach without a stopover in Hawaii.

Tuesday, January 6

We spent a little over 24 hours in Nagoya. Yesterday I took a tour of the Noritake china factory. It makes some of the best china in the world. The manufacturing techniques are unique and evidently secret, so no pictures were allowed inside the plant. That was a challenge to everyone who would normally not even think to snap pictures. Now we all tried sneaking photos in at every moment, including me. It was such a quiet factory, you could hear the click of our camera shutters. There were about 1000 girls working in this long, low-ceiling wooden-frame building with a corrugated metal roof. It was dingy-looking, was poorly lit, and resembled a giant kiln operation of a hobby store.

We also visited some famous Japanese shrines and saw many Japanese tourists all dressed up in their best and most colorful kimonos. Westernization hasn't affected Japanese traditions, especially among the women. Some of the women in their spotless outfits would pose for pictures, and some wouldn't. Before allowing pictures to be taken, mothers would straighten their kids' clothes, tear off loose ends of threads, and wipe the kids' noses. Taking pictures of the preparation for the picture taking made for better photos than the plain posing.

At a large department store I found out how to babysit the kids while parents shop. On the store's roof there's a junior-size amusement park including lots of rides. There's also a chain-link fence around the perimeter of the roof to keep the kids from falling over the edge.

Along with a couple of other guys from the ship, I climbed halfway up an observation tower to the 300-foot level. It is the second-highest TV antenna in the world. A producer showed us around the station. After the tour he guided us on a six-block walk in bitter cold to a nice, cozy Japanese restaurant for dinner, although he did not stay. Our dinner consisted of steamed rice, boiled beef in some kind of soup with onion greens, and some greasy fried shrimp. He was a very thoughtful guy. I wonder how many Americans would walk six blocks out of their way, in frigid weather to help a foreigner find a good place to eat.

As a payback, our TV station tour guide came aboard the ship today. I took him for a tour of our floating city and let him sit in the helo pilot seat as I explained how it worked. I introduced him to the admiral who was holding a reception on board the ship.

The crew turned in a good day's work on the helo. It's looking in much better condition.

This afternoon we left port and headed for Osaka. We'll arrive there in the morning. This is like a yachting tour of Japan.

Thursday, January 8

As the ship was about to enter port at Osaka late this morning, I flew the admiral off to Kobe. He made some official calls in that city, 20 miles from where the ship was to anchor. I waited at the heliport and just before dark we flew back to the ship.

The admiral was in his usual good spirits and I let him do a lot of flying including climbing, gliding, turning, and coming to a hover. He's getting very good. I asked him if I could get a picture with him on our next flight.

"No problem."

When we landed, he ordered his aide to have the photographer waiting.

Friday, January 9

I played a PIO role today, and it almost cost me my life. The ship left Osaka at 0900 hours with 75 Japanese guests and 25 members of the press aboard. We proceeded to a small island in the harbor area, picked up some schoolchildren, and then went to Kobe.

The ship's company had donated several thousand dollars to rebuild the kids' dilapidated school building. There was a special presentation on board the ship. The admiral presented plaques and certificates and gave a speech. The press had a field day scurrying around taking dozens of pictures. Those Japanese photographers will snap a picture of anything.

On the way to the island I escorted guests on a tour of the ship until flight quarters. All the guests were cleared from the fantail and herded amidship to watch the helicopter airshow. It was fun doing my usual stunts lifting off to a hover, nosing over and then just as soon as the helo cleared the edge of the ship, pushing the nose of the helo down until the landing gear was about 2 feet off the water. It's not a particularly dangerous maneuver, but the visual effects from the ship are spectacular. It appears the helo is going in the drink. Everyone gasps and some even run to the side of the ship to be sure we're still flying. It's even more spectacular when the same stunt is done from the flight deck of the carrier.

Another stunt included flying in the direction the ship was heading, then a reversal coming back at the ship, just about head-on. About the time I reached the bow of the ship, I peeled off to the right, always to the right, so I can keep the bridge in sight. This always gets the bridge excited wondering whether I'll make the turn away in time. The guys in CIC tell me it totally blots out their radar screen. That should give today's visitors in CIC something to look at.

Then I flew around the stern of the ship and back up along the starboard side. The ship was doing about 15 knots. I took a position alongside flying in the same direction. Then my crewman Pobst got into the rescue sling and opened the rescue door, and I lowered him by hoist so he dangled about 15 feet below the helicopter—just above the icy waters. It appeared we were standing still because we were flying in formation on the ship, although our groundspeed was about 15 knots.

Once I got Pobst back into the helo and the hatch secured, I turned sideways and picked up a planeguard position alongside the ship. Then I made circles to the right and to the left, maintaining formation alongside the ship. This caused a lot of pressure and force and the helo buffeted quite a bit.

Next I performed a tight series of reversal of directions or wingover-type maneuvers. It was fun putting the helo through its paces. Once the show was over I circled around the ship, then casually glanced at my transmission pressure and temperature gauges. The aft transmission oil pressure gauge was considerably lower than the one in the front.

"That's funny," I thought, and proceeded for a normal landing.

My landing was routine except for the crew signaling me to shut down fast, which I did. Then there was a big scramble among the guys around the flight deck. The ship's company spread out a giant tarp. Dunlap approached me.

"There's a big oil leak somewhere in the aft end of the helo and it's spilling oil all over the wooden deck."

"Oh boy," I thought. "That's really gonna make the ship happy."

As I took off my helmet and gloves, Dunlap came back.

"We found the problem. The aft transmission line has ruptured and oil is all over the bottom of the helo. You probably had another minute of flying before the transmission would have frozen."

Once it froze, no one had to tell me that the rotors would have snapped off and I'd have ended up in the drink.

As I ducked out the helo door, I reached into my ankle-high flight suit pocket, pulled out my fore/aft hat, and calmly adjusted it on my head before examining the leak.

The Third Division leather-faced, tough first-class boatswain's mate, in charge of the beautiful bleached wooden decks, charged over to have a word with me. He had exposed tatoos up and down his forearms featuring roses, anchors, naked women, and various graffiti. He looked like a typical sailor with his white T-shirt and his blue dungarees.

I'd watched him many times before as he harshly barked orders liberally salted with profanity to his men. I could see he wasn't very happy. In fact, he wasn't happy at all over the oil I had spilled all over his wooden decks. He sputtered in an accusing manner:

"You and your f——— helo have just spilled f——— oil all over my f——— wooden decks. Now who the f—- is going to clean up this f——— mess?"

Stunned by his comments, I gave him an icy stare, and thought that's no way for him to talk to an officer. No matter how bad the situation was, this was insubordination. I stood erect and glared at him straight in the eye and the only thing I could say in language he'd understand was:

"I just about got f——— killed out there just now when that f——— oil line broke and the f——— oil spilled all over the f——— helo, so I don't f——— care about your f——— deck and the f——— oil on your f——— deck because I'm lucky to be alive."

His eyes bugged out in quiet astonishment. He'd never heard me talk that way.

"You better f——— apologize, sailor, or I'll put you on report."

"I'm sorry, Sir, I guess I just got a little excited."

He immediately turned around and without profanity ordered his men to mop up the sloppy oil staining his precious white wooden decks. I turned and headed for the bridge where I'd been summoned.

"They'll probably be holystoning those decks for days," I thought to myself, "and with an attitude like that I really don't care."

The XO called me to the bridge for a rare "Well done."

No one mentioned the oil spillage.

That'll be the last show I do like that. I tried to figure out what caused the tubing to burst. I think all the tight turning put too much pressure on the oil lines, causing them to split, although we never did figure out the exact cause.

Statistics show the most dangerous period in a military pilot's life to be when he's logged about 500 flight hours. That's when he knows it all. I've got in my 500 hours this month. I'd better be careful.

The crew is replacing the line, and we'll fly a test hop in the morning.

Once we anchored in the harbor at Kobe, I went ashore for one of those famous fattened and hand-massaged beefsteaks. It was good, but no better than I had tasted in the States.

Kobe is the number one seaport of Japan handling 40 percent of Japan's exports and 20 percent of the imports. It's home of the Takarazuka Dance Theater, the most famous in Japan. It features continuous music and lavish Las Vegas–type costumes, only with nonsexy scenes where the music is the message. It stars all women, even the roles of men are played by women.

John Loomis and I'd seen the sister theater in Tokyo. They have four troupes. One performs at each theater, and two prepare for the next production. The productions rotate monthly.

Sunday, January 11

The ship had a local minister in for church services today. He's a Canadian missionary and is excellent. The word of Christ gets around much better out here than I thought. Evidently most of the better chaplains are out here rather than in the States.

We leave Kobe in the morning for operations at sea for several days. Then we'll go to Sasebo, Kagoshima, Moji, and Beppo—all in southern Japan.

Monday, January 12

Under way this morning, there are 36 more days before the 7-month cruise is over. Being on a cruise really teaches you how to count backward. Tomorrow we join up with the *Bonhomme Richard* for several days of defensive exercises.

Fighters and bombers from other carriers will try to attack our task group. The aircraft from the *Bonnie Dick* will make every effort to intercept the "enemy" before they can "sink" the carrier or the *Helena*. The *Helena* will be at Condition Three as our AAA gunners prepare to defend our ship. I'm scheduled to take the admiral to the carrier. He wants to watch the action of the war games from that vantage point.

I test-flew my HUP today. The oil leak is fixed. I landed seven times to keep my reflexes sharp. I shot several practice autorotations simulating an engine failure, taking them right down to the flight deck before waving off.

It's hairy coming down from 500 feet and trying to hit that small cruiser flight deck without an engine to help control the aircraft. Of course, the practice effort was controlled.

It would be rare to actually be in a perfect position to have an engine failure that would allow such maneuvering. Since it takes 15 seconds from 500 feet to get down to the water, I doubt the flight deck area could be cleared. I'd have to radio the ship on the way down, telling them the problem and my intentions. Then they'd have to call over the 1MC so anyone near the flight deck could scramble out of the way before I arrived.

One slight screwup in my technique could mess up the landing, causing a crash and probably a fire. What a mess that would be just to try to save one helo.

I've pretty well come to the conclusion that it would be better to ditch in the sea alongside or slightly in front of the cruiser if I had to.

Because the water is below 60°F, SOP (standard operating procedure) is to wear a rubber poopy or exposure suit that would keep the water from freezing me. It supposedly gives a pilot 30 minutes before dying from the cold water, although I would probably lose my uncovered hands. The suit has tight seals at the wrists, but the hands are exposed. The same type of seal is around the neck. The theory is that the Mae West should keep my head out of the water. My feet are in rubber boots that are stuffed inside your flying boots.

It takes about 15 minutes to get into special underwear and crawl through a hole in the back of the suit that is later sealed.

At the end of a flight I'm dripping sweat inside and need a shower. A lot of the guys don't wear the clumsy tailor-made suit, but I usually am flying so far from the ship it would take too long for anyone to reach me if I went into the water. So I do wear it.

My idea to stand watch in the message center has failed. I get a lot of ribbing from the officers on the ship for just flying and maintaining the helo, but they won't let me stand watches because it would keep the ship's officers from getting checked out at the various duty stations. It is more enjoyable to be busy than idle—it makes the time pass faster.

Tuesday, January 13

Mail call. Mail service has improved. It comes from the States to Japan, from Japan by COD to the *Bonnie Dick*, and I bring it by helo to the *Helena*. The seas are extremely rough today with lots of wind. The usual problems with the bridge of getting the right launch conditions were repeated.

At the start of the cruise they didn't realize how delicate the helo is, so they wouldn't give the conditions I needed. Now they think I'm such a good pilot they don't need to give me the right wind and deck conditions. They just don't realize I have to baby these birds to keep from banging them up.

Today I flew the admiral and Vice Admiral Terai of the Japanese Navy, who is a guest of the admiral, over to the carrier and back. The ship is rocking and rolling tonight. The crew has the helo secured very tight. Corrosion will be a big problem, since there is lots of saltwater spray tonight.

So far this cruise, no seasickness, but I've come close a couple of times. The best prevention I find is to have a full stomach.

Wednesday, January 14

At 0530 this morning I was awakened by a messenger from the OOD. While getting planes ready for launch, a plane captain on the *Bonnie Dick* had walked over the edge of the flight deck in the dark at 0500 and fell into the sea. They wanted to launch me in search of him as soon as possible so the *Bonnie Dick* helo could operate planeguard duties. This would allow flight ops and the exercise against the "enemy" to continue.

I told the messenger to go back and rack out the crew. I figured we could get airborne at 0650 hours, 10 minutes before sunrise, and they needed to hurry preparations for launch. The USS *Brush* (DD-745), a destroyer, was taking up station about where the man had fallen over. The man overboard Oscar flag flew from the yardarm.

I stuffed down some breakfast and plenty of juice and water. We launched right on time with 30-knot winds, rough seas, and rainshowers. I started several search patterns using the destroyer as a guide. I carried enough fuel for an hour and a half of flying. I didn't want to overload the helo in case we found the victim; we'd be able to hoist him aboard.

In the morning I flew 5.4 hours, landing only to refuel on the run, but not shutting down the engine. The hours consisted of continuous searching at low altitudes. The whitecaps were whipped up by the winds so that I couldn't see anything on the surface at 500 feet. I changed crewmen each time I landed. It was rough flying with lots of stick movement correcting for all the rough stormy air and winds.

I landed for lunch at 1230 for an hour's rest. I flew another 3.6 hours searching in the afternoon for a total of 9.0 hours with only five landings. It was an awful lot of flying and all with negative results.

During the morning flight I had a terrible problem. All the juices and water I'd had for breakfast filled my bladder. I had to take a leak something terrible. There is no relief tube in the helo, so on the first fuel stop I had Dunlap get me a glass jar while the crew was refueling the helo. I was too embarrassed to use it in front of everyone peering into the helo as we sat on the flight deck.

Once I got airborne I tried to use it. What a mess trying to hold the bottle between my legs, unzipping my flight suit, using the bottle, and flying the helo with both hands, then trying to dump the contents out the side window, but keeping the jar in case I needed it again. Fortunately there was time for everything to dry before my next landing.

It's really hard to determine how long to search for a man overboard. In seas like today I think he could have lasted no more than 5 or 10 minutes. But if I didn't do everything possible to save a shipmate, I wouldn't feel right. My occupation is saving lives regardless of the odds, so I wanted to make every effort possible.

The admiral in charge of the carrier sent a nice message to the captain of the *Helena* about my search efforts today.

FROM: **CTG 77.5 (CTG = Commander Task Group 77.5)**

TO: **USS BRUSH (Destroyer who helped)**

INFO: **COMDESDIV232 USS BONHOMME RICHARD AND COM-SEVENTHFLT**

ORIGINATOR JOINS BONHOMME RICHARD AND AIR GROUP 19 IN EXPRESSING APPRECIATION FOR YOUR CONSCIENTIOUS AND THOROUGH SEARCH FOR OUR UNFORTUNATE SHIPMATE X WE SINCERELY REGRET THE NEGATIVE RESULTS BUT YOUR EFFORTS MADE POSSIBLE PARTIAL COMPLETION SCHEDULED

ADEX X IT MUST HAVE BEEN ROUGH ON THE HELO PILOT X HE MUST BE A TIGER X WELL DONE ALL HANDS

Thursday, January 15

After the message from the *Bonnie Dick* was circulated, I received a new nickname, "Tiger." It replaced a few others like "Zoomie," a name I got for zooming the ship to tease the blackshoes on how close I could come, or "Beanie," which referred to the skullcap with the propeller on top, plus a couple of others like "Dangerous Dan" and "Deacon Dan."

I made a short trip to the carrier today to pick up a new oil tank from their supply system. We developed a leak in our oil tank, but the crew plugged it for the trip over and back from the carrier. They installed the new one, and we're back in full operation.

As we were being refueled by the tanker this afternoon, the admiral highlined over to visit the crew on the oiler. He gets around to see all the ships under his command.

I think I've worked out a deal to stand OOD (officer of the deck) under instruction watches with my goal toward being qualified as an underway OOD for independent steaming by the time we get back to the States. That means the captain would trust me to run the ship while steaming alone out at sea. An important qualification for a young naval officer.

Friday, January 16

We've arrived in Kagoshima. Tomorrow I'm scheduled to fly around to various airfields in the vicinity of Kagoshima for an inspection tour. But will the weather cooperate? This morning it rained, this afternoon it snowed, and tonight it hailed.

I planned to go sight-seeing this afternoon, but with the rain and snow it wouldn't be good for taking pictures, so I took a nap. We had a flat tire on the helo tailwheel this morning. The crew fixed it. The saltwater is causing us tremendous corrosion problems.

Saturday, January 17

The snow kept us out of Kagoshima. This morning it was snowing so hard the visibility was several hundred yards much of the time. Naturally there was no flying. Since the admiral's schedule is built around helicopter travel, it shot down all the events he had planned, so we left port 12 hours early and headed for Sasebo.

Seas are rough, and it's cold out tonight. The ship is bouncing around like a chip of wood on the heavy seas.

The admiral ate with the officers in the wardroom tonight. There are two sittings—one for junior officers and one for senior officers. He ate with the seniors, and I was invited to sit at the head table, right across from the admiral. I was flattered to be included.

Sunday, January 18

The ship arrived at Sasebo this morning. It was snowing heavily, but stopped in the afternoon. The temperature is in the 20s. We should get some snow on our cross-country tomorrow. I've never had to follow maps where all the landmarks are covered with a white blanket.

Wednesday, January 21

We're tied to a pier at Moji. It's a giant wharf with a two-story cargo building on top of the pier. There are moving cranes that run up and down the wharf on a pair of railroad tracks.

The weather remains lousy—biting cold, low overcast with strong gusty winds. These are marginal conditions, but okay to fly.

The American Consul of southern Japan, a man named Herndon, wanted to go to Itazuke Air Force Base, about 45 miles away, and later I was to bring the admiral back from there. We climbed in the helo, cranked up, and lifted off. The wind was hitting the ship from the bayside and pushing it against the pier. It was perfect for takeoff right into the wind. As we climbed I gave a passing thought to what wind conditions would be for landing on the return, and then focused on navigating across country to Itazuke.

As we headed out, I followed the line drawn on my kneeboard map matching the features with the terrain. Occasional railroad tracks and roads were great checkpoints for verifying the course on the cross-country trek. Because of the strong winds, it took a 15- to 20-degree crab or heading off course, to remain on track across the ground and on my timeline for the whole trip.

In a strange country with mile after mile of flat rice paddies and agricultural acreage, with limited visibility due to the low scuddy clouds, the pit of my stomach was constantly knotted. I wasn't positive that I wasn't lost. Outwardly my passenger had no inkling of my concern, but it was there.

We reached Itazuke with no problem. I refueled and prepared my flight plan for the return. The admiral showed up on time for the return trip and had a captain in tow. I let the admiral fly a great portion of the trip.

When we leveled out at 500 feet, I handed my 8mm movie camera to the captain in back, and since he didn't have a helmet on, motioned him to take a picture of us in the cockpit. I did a bit of comedy waving my hands in the air with exaggerated lip movement saying "Look mom, no hands." Then I raised my feet high for the camera and mouthed, "Look mom, no feet," and then with an open mouth and lips over my teeth, "Look mom, no teeth." All the time the admiral is working all the controls. He's trying to concentrate and laugh at the same time. I wondered how many jg's would try that with a three-star? He seemed to enjoy it.

It was time to get down to business having him climb and glide, do turns both right and left, and even do a couple of 360s. His coordination is getting better, including controlling the rpm.

When he got fatigued, I took over. We talked about a lot of little things and then he said over the intercom, "Lieutenant, pretty soon the *Rochester* is going to relieve the *Helena* as flagship. I'd like you to move over with my staff and become my permanent personal helicopter pilot."

My heart sank. The staff had approached me about this possibility, but this was the admiral himself. What do I say to the commander of the most powerful naval force ever assembled in the history of the world? Here I am just a kid, and he's asking me to become his personal pilot. It would guarantee a good fitness report, great career opportunities, and travel to every major area in the Far East.

While I was trying to think what to say, he continued, "I understand you've been trying to stand watch on the ship, but it hasn't worked out. When you move to the *Rochester*, I think I can help you get to stand watch on the ship and if not there, then we'll find a slot on my staff. How about it?"

I wanted to go home.

"Admiral, I've got my wife and everything back in the States and I just want to get back there."

"I understand that," he responded.

"We have the staff wives living in Yokosuka, and I could make arrangements for her to come over to Japan. Then when you're in port you could be with her."

I looked out the window and then at my kneeboard checking the maps—trying to look like I was busy navigating. Heck, my mind was blank and nothing I looked at registered. I was really trying to figure out what to answer.

"Admiral," I finally stammered, "I'm really flattered by your suggestion, but I've been out here 6 months and I'd just like to get back to the States. My plans are to leave active duty when my tour is up the end of this year. The squadron has another pilot on the way out here, I don't know what else to say. I'm so honored by such a request, Sir."

Silence.

Then, "Well, Lieutenant, my staff told me, but I thought I'd try. I enjoy flying with you and like you around."

"Thank you, Sir," and I motioned for him to take over the controls again.

I made the approach to the ship. The 45-minute hop had taken an hour because of the admiral's extra maneuvering. The wind had picked up and was really strong. The only way to approach the ship for a landing was downwind from the bayside. The cranes were obstructions and the cargo storage structures built on the pier made an into-the-wind landing approach impossible. I hovered, facing downwind alongside the ship, but the strong winds from abeam required full-back stick. There was no room in the cyclic for maneuvering, so I did a 180 and asked the admiral over the intercom to look out his side window for the crane on the stern of the ship.

I looked out my window and let the wind gradually drift the helo back toward the flight deck. My crew got the idea, and Dunlop went forward along the rail of the ship to try to signal me. I was purposely high and edged back slowly. The admiral kept saying, "You're clear, you're clear."

I had to be careful. If I went too far back, the crane on the dock along-side the ship could knock the rear rotor blades off. Soon the edge of the ship was directly below me. I gradually settled down and took my proper position on the flight deck and landed. It worked out fine.

"That's the first time I ever backed into a landing," the admiral exclaimed over the intercom.

It was mine, too.

The staff told me later he couldn't stop joking about backing into a landing.

Sunday, January 25

We're visiting Beppu—our last stop on the admiral's goodwill tour of Japanese ports. The admiral was flown to a nearby airport to catch a COD to a carrier that will rendezvous with us tomorrow on the way back to Yokosuka.

We'll be in Yokosuka for 9 days while the admiral's staff is transferred to our relief cruiser, the USS *Rochester* (CA-124), and then the *Helena* will head for the States with me on board. The helicopter will be left behind at Oppama.

There's an abandoned airfield near where we're anchored. As we took off this morning, I asked the admiral if he'd like to shoot an autorotation. He seemed in good spirits even though he was out late last night at a reception and dinner. He said, "Sure, let's do it."

I don't think he knew what he was getting into. For safety sake I climbed to 600 feet to be sure we had plenty of autorotation space. I asked him to strap in tight and indicated the same to his commander aide in the back.

I planned a 90-degree autorotation and for the second one I'd give him the full treatment with an exciting 180.

We came over the field. The wind was calm.

"See the X. That's our spot we plan to land on," I told him on the intercom.

The commander behind me was listening on his helmet headset and strained over my shoulders to see what I was talking about. He could tell, by my careful preparation and precise approach pattern, that this wasn't an ordinary event. Being an aide to an admiral certainly can't be the best job one would want because you're just along for the ride. He had no choice in this matter.

"All right, here we go, hold on," I warned as we hit the 90.

I pushed the collective down, and we instantly started to sink at 2000 feet per minute. I gave it left cyclic and rudder and dipped the nose to commence a left 90-degree turn and maintained airspeed at 60 knots. I had the collective full down—to the bottom to get the hollow wooden rotors up to speed.

Like all HUP autorotations it felt like the bottom fell out and we were free-falling toward the ground.

As rotor rpm built up, and to keep it from overspeeding, I gradually pulled up the collective so that the needles stayed within the correct parameters.

The admiral was thrown out of his seat, barely restrained by the loose seatbelts and shoulder harness. He hadn't really tightened them.

He grabbed for the window ledge. Behind me the aide was grabbing for anything he could get his hands on. He had no shoulder harness.

After the initial shock, we settled down and sank as the rotor blades windmilled.

At 100 feet I brought the nose up, slowed airspeed to 45 knots, dug into a flare at 50 feet, and immediately leveled off. Simultaneously I pulled up on the collective to marry the engine and rotor blade rpm needles. This indicated that the engine and the rotor blades were connected.

We scooted in a forward motion and started a climb for the 180 auto—the real Disneyland E-ticket ride.

As a courtesy on the climbout, I asked the admiral if he'd like to try one, knowing full well he'd never take me up on the offer.

"Lieutenant," he responded, "I had a rough night last night. I think we better just head to our destination. That was a very interesting maneuver."

"Yes, Sir."

I could feel the commander's tenseness. He went limp with relief that we were knocking off the autorotations.

The old jet jocks aren't as fearless as they'd have you believe.

Monday, January 26

My last operational hop from the *Helena* took place today. I flew to the carrier and brought Admiral Kivette back to his flagship for his last day at sea aboard the *Helena*. Vice Admiral Rubin Libby, commander of the First Fleet, came along. The commander of the First Fleet was in charge of all Navy ships based along the West Coast of the United States.

Later in the day Admiral Kivette posed for pictures with my helicopter crew. I thought each crewman would enjoy having his picture with the admiral.

While we were standing around, he said, "Lieutenant, you've got one last chance to join me. We can still make it happen even at this late date."

"Thank you very much, admiral, but I'm going back to the States on the *Helena*."

We shook hands, and he returned to his flag quarters.

I thoroughly enjoyed flying for him.

While we steamed for Yokosuka, the XO called me to the bridge. The captain was with him.

"Mr. McKinnon, I'm very happy with your performance as helicopter pilot on this cruise. You did a fine job."

"Thank you, captain."

"We're proud of your safety record, Lieutenant," added the XO.

So was I.

It was somewhat of a lovefest, patting each other on the back about how great the helo operations had gone the last 7 months and how we'd worked together as a team. We had actually done that after all the fights.

The heated words, bickering, and fighting about operational conditions were forgotten. We were all glad we had completed it safely. A pilot can get good at shipboard operations, but he always wonders when it might be his turn for something bad to happen. There's always that danger.

We did have one of the most successful cruises ever. Plenty of rescues (60), record number of flight hours (211.4), record number of flights (235) with 442 shipboard landings, and 152 shore landings. We also had a perfect safety record.

Every other cruiser pilot I've heard about out here has had some kind of accident—smashed tailwheels, busted rotor blades, and even engine failures. Considering all the hazards involved, I had been very fortunate. I thank God for the perfect safety record. There had been too many chances and near misses. Without his protection it would have been impossible to have ended up unscathed.

In the afternoon as we entered Tokyo Bay approaching Yokosuka, I launched for my last hop to Oppama.

I buzzed the ship several times in a farewell salute, and all hands waved back. There's a real euphoria that sets in at the end of a cruise. Everyone is glad it's over. Everyone has fond memories, a confident feeling of being among the best, but yet a sincere desire to get home.

Tuesday, January 27

Det One made an end-of-the-cruise inspection of the helo. They said it was the best kept airplane they'd seen any ship return to them. The crew had done an outstanding job of maintaining the bird. They will receive very good marks on their fitness reports from me.

Lieutenant jg Bud Smith is relieving me. I have been pumping him full of information. He's new to cruiser operations, and I'm happy to be able to help him, particularly since I had such a lousy briefing and learned everything the hard way.

All the extra parts and VIP gear we've accumulated the past 7 months were turned over to him.

Today's my birthday—25 years old. I don't feel much older or weaker, nor do I see additional gray hairs. What a great birthday present—to have this cruise over safely.

Tonight the officers of the *Helena* hosted a get-together for the officers of the *Rochester* and the Seventh Fleet staff at the "O" Club.

I introduced Bud to everyone, including the admiral.

The admiral spent a disproportionate time with me and suggested I bring prints of the pictures taken of us together yesterday so he could autograph them.

"Yes, Sir."

I asked the CO and XO of the *Helena* to speak to the CO and XO of the *Rochester* about helo ops to help Bud. They agreed to do that. Bud doesn't seem too aggressive or sure of himself. I hope he gets some self-confidence quick, or he's going to be in for a rough cruise.

A lot of the sailors on the ship have made flattering comments to me. I think there are two reasons for the attention. First, it always helps to have a smile. Second, you have to be somewhat of a character to stand out. It isn't much trouble for pilots to stand out from the blackshoes.

We're leaving the helo in Japan. That means that when we arrive at Ream Field our logistics will be easy—just get our personal gear off the ship.

Wednesday, January 28

All my business with *Det One* is completed. Now I just sit around and wait for the admiral's staff to transfer all their files and gear from the *Helena* to the *Rochester*. When that's complete, we'll depart for the States. Meanwhile I'll stay at Oppama, get in a little flight time for me and the crew, take some pictures of Mt. Fujiama, and do some final shopping at the exchange, as well as a little more sight-seeing. As a "has been," I am anxious to return to CONUS.

Thursday, January 29

The ship pulls out 4 February at 1600. There seems to be a "little" discussion about Bud relieving me. The scuttlebutt is that someone is raising questions about his ability to handle the aircraft aboard the cruiser since he's never done it. I'm lying low and am anxious to get aboard the *Helena*, and head out of here before anyone gets any fancy ideas.

Friday, January 30

I had a flight physical at the Oppama Medical Center. I'm in top shape, but in lousy condition from sitting around for 7 months without much exercise.

I stopped by to get the admiral's autographs on the photographs for the crew. He asked me to pay his respects to Admiral Pride, the top admiral in San Diego.

I managed to get in 4 hours of flight time on Monday and Tuesday.

Monday, February 2

Bud Smith and I put in an hour of flying time in the HUP assigned to him. The doggone thing practically fell apart while we were in the air. We nursed it back to Oppama. It had some unusual vibrations. *Det One* wants to keep my UP-18 and stick Bud with this piece of junk. I think they'll change their minds because the helo we flew today needs a long time of testing and fixing to make it fly correctly.

I attempted to work on my cruise report this afternoon, but it's very difficult to settle down and concentrate. Maybe I'll do it on the ship on the way home. Everyone associated with the ship after 7 months is so excited about getting home it's hard to sit still.

Tuesday, February 3

We cast off lines at 1600 tomorrow. The lousy, foggy weather kept me from flying and giving some of the people from the ship a helo ride. Now that the cruise is over, they want to fly in a helo; they were always chicken before. Until about a year ago pilots finishing a cruise had the privilege of flying home aboard a space-available MATS (Military Air Transport Service) flight. Budget cuts now dictate that I ride home with the ship.

Tuesday, February 10

A truce was struck. We're under way home, but the ship wouldn't let me stand OOD watches as promised. They needed the training time for their own ship's company, so I started sacking out during the day like I was a guest on a fancy cruise ship. The gunnery officer didn't like that one bit even though I don't have any responsibilities. He ordered me to stand watches in the steaming-hot engine room. There wasn't much point for a naval aviator to know about steam engines, supersaturated steam boilers and stuff like that, so we struck a deal. I wouldn't sleep in the daytime if I didn't have to stand watches in the engine room.

But I did learn an interesting tidbit. The engine annunciator indicators show speeds using words like "slow," "standard," and "flank." With each indication the propeller shafts are supposed to make a certain number of turns that controls the speed of the ship by controlling the number of rotations of the propeller.

"Channel fever" has reached epidemic proportions, and the guys in the engine room have been putting on extra turns for our nonstop journey to Long Beach. The navigator took a star fix last night and discovered we've been covering miles through the ocean at a faster clip than we should. Matter of fact, we're almost a half a day ahead of our scheduled arrival time in Long Beach. Navy ships don't like to arrive early.

The engineering officer got chewed out because of this. He chewed out his snipes, and now we're moving very slowly to get back on our schedule. Word of this has buzzed all over the ship, and the morale sank. The captain and the navigator didn't win any friends with this episode.

After 12 days under way since leaving Yokosuka, we finally arrived in Long Beach 215 days after we left. It's great to be back in America.

CHAPTER TWENTY-ONE

Full Cycle

T HERE'S A NATURAL LETDOWN returning to the squadron. You're happy to be back from cruise, but you miss all the action. It's a bittersweet feeling. You like being home, but you like the flying and the constant test of your skill and abilities, too.

The purpose of the U.S. Navy is to provide a fighting force to back up America's foreign policy abroad. If you're not aboard ship on cruise as part of that effort, you don't feel like you're part of the main team. I'd had the opportunity to land on every cruiser and carrier assigned to the Pacific fleet.

As someone who decided not to remain on active duty after my required 2-year commitment, much of the excitement fades after completing a long cruise. Returning to the squadron, I was assigned to my old slot as PIO, and designated as one of several squadron test pilots, I also served as an instructor pilot to shepherd along newly reporting HUP pilots preparing for their first long cruise. It's an ever-evolving squadron training cycle just like I'd been through one short year ago.

One of the toughest things to adjust to on returning from a cruise is that you don't fly alone anymore.

For 7 months I owned my own helo and almost never flew a dual-time flight. Budget constraints have come into play, and to make maximum utilization of aircraft and pilots, all flights except test flights, are dual with the splitting of the flight hours between the pilots. Fortunately my experience as a cruiser pilot puts me in demand, so I get plenty of flight time.

One of the most interesting hops here at the squadron included a network television stint with Jack Linkletter (Art's son), demonstrating rescue techniques from a carrier moored in San Diego Bay. Jack sat in the back of the helo as we took off. We circled near the carrier and hoisted a crewman out of the bay while the network cameras on the ship recorded the whole event.

Our new XO, Lt. Commander Bob Shields, is the first Navy man ever to be rescued by a helicopter. He had an engine failure after takeoff in his SB2C dive bomber after leaving the deck of the carrier USS *Leyte* (CV-32) on February 9, 1947. Chief Sikorsky test pilot Jimmy Viner, who was performing some early helicopter demonstration flights, plucked him from the water before the oncoming carrier could run him down.

I made quite a few one-week training cruises for new pilots to prepare them for their long cruises.

One such cruise was aboard the USS *Oriskany* (CVA-34), on July 26, 1959. The ship was just out of overhaul. As yet no planes had landed on the new deck while the ship was under way. To be the first is a big deal to aviators. I ferried our planeguard HUP aboard the ship as it was under way pulling out of San Diego Harbor.

The CAG exploded.

He was scheduled to make the first landing aboard the modernized ship as soon as we got to sea.

Oddly enough, a similar incident occurred on the original flight deck landing aboard the *Bennington*.

Author Tommy Thomason relates, "the first helicopter landing on the same carrier was extemporaneous and not well received. Helicopters became popular in 1944 for calibration of ship's radars since they could hold a position at bearing, range, and altitude. Lieutenant jg Barney Mazonson was assigned to fly an HNS as a calibration target for the fire control radars on the newly commissioned *Bennington* (CVA-20).

"Due to the lack of performance of the HNS, this was normally accomplished solo with a radio installed. Mazonson somehow wound up in the training configuration with the radio replaced by a student. Since he needed instructions about position requirements, he nonchalantly landed aboard.

"Unfortunately, nobody had bothered to inform the captain or air operations officer that not only a landing, but the momentous first landing, was about to occur. They were outraged. Mazonson was finally permitted to leave after the flight deck was cleared for flight operations. He took off, backward, probably grateful that the fire control radar was not yet calibrated."

During the week I got a hop aboard an AD-5N and made two arrested landings. I was sure we were going to end up ripping up the flight deck with the blurred disk of the prop the way we bounced forward after the tailhook caught.

The relief cruiser for the *Rochester* was the USS *Bremerton* (CA-130). It was taking a 2-week VIP training cruise from Long Beach to the Portland Rose Festival in June. I was assigned to head the detachment with Lieutenant Russ Ackley, the pilot assigned to the cruiser for the long cruise. The squadron thought we would have an excellent opportunity to get some training time for Ackley and to train the ship in helo ops. What a farce.

We managed only a total of 2.6 flight hours with three flights and four shipboard landings during the 2 weeks. The captain on the *Bremerton* was worse than *Helena*'s captain. He wasn't interested in helo operations or learning some of the finer points that would help him down the road when they deployed. It was a great opportunity to train, and it was all wasted. To the blackshoes and the aircraft carrier brownshoes the only good helo is a parked helo—unless you're the man overboard or the aircraft that goes in the water.

The squadron has been evaluating a couple of scoop nets with retriever poles for picking up unconscious rescuees. The general feeling is that the rescue sling with the Chicago grip is still the best method when a crewman enters the water.

Soon we are to receive about 30 three-prong rescue seats which will be standard equipment for conscious pickups.

Safety has been a constant concern. There have been 40 major helicopter accidents per 100,000 hours flying in 1959.

We've had a tough year at HU-1.

While flying from nearly every type of ship in the U.S. Fleet and from most ports in the Pacific Ocean, HU-1 flew more than 15,000 hours and made about 11,000 shipboard landings, and heaven only knows how many field landings. This is quite a bit of chopper flying when you consider that other West Coast helo squadrons average about 4000 hours a year.

On the downside, we had 10 major accidents and one minor accident—about double the Navywide average. This was more than all the other Ream Field squadrons combined. Our squadron goal was 2.9 per 100,000 hours.

There were many comments about how many men HU-1 had deployed on TAD (temporary additional duty) orders. Our PIO department found in the last year the squadron made 183 sets of TAD orders for a total of 384 officers. It amounted to more than 14,800 days or 40 years of TAD time. Combined with the enlisted men's orders, the squadron could send one man on temporary additional duty for a period of $135^{1}/_{2}$ years. That's a lot of away-from-home time.

CHAPTER TWENTY-TWO

Final Rescue

O N SEPTEMBER 20TH I deployed as O-in-C of Unit 55 aboard the USS *Ranger* (CVA-61). It was my last cruise before leaving active duty in December. This was scheduled as another routine training cruise for 2 weeks to train some pilots and enlisted crew.

It was also going to end up as a disaster.

The ship was to operate out of San Francisco, where it was homeported. Since HU-1 was the only planeguard squadron on the West Coast, we ferried the two HUPs from San Diego to the Oakland Naval Air Station to join the ship. It was a 3-day, 8-hour-plus cross-country. There were two pilots in each helo, and as we flew in formation we shared the flying. The enlisted crew flew up by Navy transport. The ferry trip was uneventful and we deployed with the air group aboard on September 21. Flight ops commenced shortly after we steamed under the majestic Golden Gate Bridge.

The early part of the first week, the jets had emergencies fairly regularly. We were operating with A4D Skyhawks and F8U Crusaders. The second week we operated with F3H Demons, A3D Skywarriors, and F8U-2 Crusaders. We've also had our share of helo mechanical problems, and the crew had to work hard the first week to have our helos in an up status. The confusion on this ship is unreal.

No one knows what's going on. Everyone wants to run the ship. There are plenty of conflicts between the CAG and the captain.

After seeing how efficient things can work on a long cruise, I guess the ship and the air group just have to go through this chaos to get ready for deployment. It is part of the process of learning how to work together. No rescues and no injuries so far.

The winds are a little fresher and more brisk in the northern California area than off the southern California coast. The seas are a little rougher, but the flying conditions are fine.

We rotated the flights among the four of us to balance the flight time. I tried to schedule the new guys so they get the maximum amount of training and experience. When not flying, I stayed in Pri-fly or in the nearby "Vulture's Row" or "Roost" on the island superstructure. One has a spectacular view of carrier landings from there. My purpose was to supervise the helo ops and be handy in case the air boss had any problems. The cruise went well.

On our final day, Thursday, October 1, we'd completed a good 2-week training. The air group had the usual series of problems and emergencies with their jets.

The ship had one last afternoon launch and bounce session scheduled, and then the air group would fly off. As the ship entered the San Francisco harbor, it was our plan to fly off both helos prior to reaching the Golden Gate and head south for San Diego. We figured we could make Ream Field by dark the next day. The crew would fly back down by transport, beating us home by 24 hours. The 16 hours of ferrying time on each helo for this cruise seemed like a lot of extra time needlessly added to the planes, but I don't know how else you'd get them up north to the San Francisco area.

It was my turn to fly the last hop of the cruise. The ship had a good session of bounces, and the planes headed off to NAS Alameda and NAS Miramar. Flying planeguard for this last session was nostalgic. I'd had some of the greatest flying experiences and responsibilities a junior officer could ever ask for. Everything LCDR Fortin had told me at the start of my long cruise had come true.

My mind was drifting to the events of the past 2 years. As I watched the launches and recoveries, I'd achieved the most cherished dream of my life—to earn a set of Navy gold wings and have as good an adventure as I'd had and to top it off with 61 rescues, a peacetime record in this business.

After the last jet launched, the *Ranger* started its turn downwind to recover me.

I left the planeguard position and started a large circle in preparation for the recovery. As my mind wondered, strangely I thought about achieving 63 rescues. I don't know why I focused on 63 rescues, but I did. Then just as strangely, something told me I'd get only 62, not 63—like a premonition. But my cruise days were over, and all the planes were gone now, "so how would I end up with any more rescues?" I thought to myself. Then I got back to business.

"Angel, Gray Eagle, we're turning downwind to recover you."

"Roger, Gray Eagle."

I moved the helo around low, zigzagging over the water killing time, but positioning myself to come aboard once the ship straightened out on course.

Then Pri-fly, with an urgency and fear in the voice of the air boss radioed, "Angel, we have an emergency. Two men overboard, starboard side. See if you can get them."

"Roger, I'll look for them."

I was astern of the ship off to the port several thousand yards. I nosed the helo over, picked up speed to get in position, and followed the wake of the ship on the starboard side.

The bridge had tried a Williamson turn, a maneuver to miss or keep from running over the men in the water. The ship was decelerating.

"The two men overboard are from your unit, Angel."

"Roger."

"How could two of my men fall overboard?" I asked myself.

Then I spotted one head bobbing in the cold water in front of me and about 50 yards further up was the second. I approached the first. R. M. Dobbins, AD2, my crewman, had the rescue hatch open. He was starting to lower the sling. The first man in the water looked like he was just treading water with no problem. Both men could clearly see me.

Then the man up ahead started waving frantically and going up and down and in and out of the water. He appeared desperate and about to drown. The man I was hovering above appeared to be okay.

Neither had on a lifejacket.

They were lined up right into the wind. It was logical to pick up the first man and then move forward to pick up the second man without any extra maneuvering. The squadron doctrine for downed men was to relax, be still if you had no problem in the water, but to wave frantically if you were in trouble.

I had a split second to make a decision.

Whom to rescue first?

It appeared that the upwind man was in deep trouble. I didn't want to be rescuing one and watching the man in front of me drown at the same time.

I overflew the first man and dragged the sling to the one waving his arms. I hovered low—real low, just a couple of feet above the water. The lower I was, the less cable had to be paid out and the quicker the pickup so I could swing around and get the second man in the water.

The first rescuee had a hard time getting into the sling, but when he was finally in, we hoisted him fast. The downwash was spraying the helo and canopy. I just held the helo in a steady hover until his head came in the hatch.

Time was urgent, and I knew it.

I pushed the left rudder to turn in the hover and did a tight 180 to come around and pick up the second man.

Meanwhile our rescuee crawled shivering to the rear of the helo and sat there.

"Gray Eagle, Angel. I have one man. Looking for the other."

"Roger, Angel."

But I couldn't see the other guy in the water. "Where was he?" I thought. My heart sank.

I slid the helo sideways to find him and frantically moved around the area at low altitude. My crewman Dobbins crawled over the rescue trapdoor and scanned out of the canopy.

This was Dobbins' second rescue. Back in April he had helped pull an AD driver from the ocean after the pilot ditched his aircraft 3 miles from the *Bonnie Dick*.

The sea was whipping up plenty of whitecaps, and the downwash of the rotors created a lot of spray. I could see nothing but dark blue ocean.

I could see no head above the water anywhere. The ocean is so massive and a human head so small.

I was frantic. When it's one of your own men, even though you don't know him, you press all the harder to be successful.

As I hovered and flew in small circles, the wind pushed on the helo. Soon I lost track of the exact spot where I had made the previous pickup. There was no head bobbing anywhere.

"Gray Eagle, Angel. Can you launch the other helo to assist me?"

"Negative, it's been damaged."

"Damaged? What could have happened?" I thought to myself.

No time now to think about that. I had Dobbins throw out a smoke flare at about 100 feet to mark where we were.

I did not want to lose sight of that spot in the ocean.

The flare failed to ignite.

They were supposed to be dropped from 200 feet to hit the water with enough impact to ignite.

I climbed up to 200 feet. The crewman dropped out another. It failed to light.

The damn flares work only about half the time under normal conditions. Now to have a series of duds when I really needed help was exasperating. I was really frustrated, and 200 feet is too high to see much in these sea conditions.

The third one finally lit, but the ocean is vast. I still wasn't sure of the exact location I had made the rescue. Precious time had been lost marking the spot. The smoke flare was approximate at very best.

I started working search patterns around the smoke flare.

Two hundred feet was too high. It was impossible to see anything the size of a head in the water. I had to go down to 50 feet, which made it easier to see, but also cut down on the area of the ocean I could search.

I continued to hunt. After 25 minutes we dropped another flare. It worked this time.

The guy in the water had disappeared. I didn't see it happen, it happened too fast. After 45 minutes the red fuel warning light came on. I spent another 10 minutes searching.

The carrier was drifting nearby at a virtual standstill. The edges of the flight deck and catwalks were lined with hundreds of men searching for the man overboard. They had launched a small boat that was crisscrossing in the rough seas searching for the man as well.

It was a futile effort.

My past search experience had taught me you had to make a rescue quickly. All my long searches in WESTPAC had been in vain. You never want to give up hope, but you've got to be realistic, too. Whoever it was, had drowned.

"Gray Eagle, Angel. I'm low on fuel. I can't find the other man. Request permission to land."

"Clear to land, Angel."

We landed and shut down. The air boss felt it was useless to continue any search. I agreed, but hated it.

"What happened?" I asked.

"We had spotted the other helo on the number one deck edge elevator at hangar deck level prior to being raised to the flight deck for launch. Just after the blueshirts cleared the elevator as we were turning downwind, a gigantic wave of rough seas cascaded over the lowered elevator like a sledgehammer. It washed your two crewmen overboard and snapped the tiedowns and overturned the helo. It was a freak wave and destroyed the helo. The helicopter is laying on its side all smashed up with the blades broken and is still dangling on the edge of the elevator.

"Two other members of your unit were injured and in a lot of pain, but managed to hang on to the helicopter and kept from being washed overboard. They are Ensign Joe Vaden and Richard Slack, AD1. Both are in sick bay."

I'd never heard of such a thing.

"One of the ship's cameramen was shooting some pictures and we may have a photo of the wave as it swept over the elevator."

The rescued crewman was Dale Hudson, AT2. The fellow that drowned was Airman Apprentice Daniel Johns.

I went to my stateroom to check the personnel records of each crewman. Johns was 20 years old, and as fate would have it, lived in Hayward, California, right outside San Francisco.

I talked with the captain, the air boss, and the chaplain. We made the decision I'd fly the remaining helo into the airfield at Oakland along with another pilot and the chaplain. The ship would make arrangements for a car, and the chaplain and I would drive to Johns's parents' home and share the bad news. It was a decision I felt was right, but later regretted.

I changed from my flight gear into my khaki uniform.

We launched about 10 miles from the Golden Gate and flew under the bridge.

It was an awesome sight—truly spectacular, but I couldn't enjoy it. My mind was on the tragedy at hand trying to figure out how to tell Johns's parents.

What do you say?

After landing, the chaplain and I climbed into the car and followed a roadmap to Hayward and Johns's home. Conversation was limited.

I was mentally reviewing my rescue sequence and techniques, probing my mind to see if I had followed all the correct procedures. I was convinced I had, but still numb from the fact that after 62 successful rescues, I had lost a man. As I relived and relived the experience in my mind, I was convinced that with the rough seas and no life saving equipment, one of the two men would have drowned. Fate had determined which one.

I was jolted out of my daydreaming as we pulled up to the curb of a small tract house on a narrow lot with not much yard.

I still didn't know what I was going to say, and I didn't think the chaplain did, either.

We knocked at the door. A young girl, about 16, answered.

"We're from the Navy. Could we come in and talk for a moment?"

"Sure," she said innocently.

We opened the screen door and entered the living room of a very average home.

As we asked if her folks were home, a middle-aged man and woman entered the room along with another young daughter.

"What's going on?" they inquired in a puzzled tone.

"Can we sit down and talk to you a moment?" the chaplain asked.

"Sure," he said, without any thought or idea of what was about to come.

The woman, obviously Daniel Johns's mother, looked a little confused as to why a couple of Navy officers were there.

"My name is Chaplain McCormick, and this is Lieutenant McKinnon," he stumbled along. "Today your son Daniel died in an accident while operating aboard the aircraft carrier *Ranger* near San Francisco."

The dad looked stunned—just like someone had punched him in the stomach and knocked all the air out.

The mother was speechless and just sat there.

The young sister, who was standing, started screaming and running around the living room and into the back of the house and back to the living room.

"Dan's dead. Dan's dead. No. No. No. He can't be. No."

Tears were in my eyes. I thought to myself, I don't know if I can handle this!

After about 10 minutes, we all managed to settle down. The chaplain offered comforting words and some thoughts about salvation from the Bible.

The dad wanted to know what happened. I tried to explain as best as I could.

Then came the second shock.

"Where's his body?" the dad asked.

"He was lost at sea," I said.

"You mean there's no body?" the dad said in a tone that almost demanded we produce his son's body.

"No, Sir, I'm sorry, he was lost at sea and we were unable to find him."

The sister again went into hysterics and raced around the house yelling, crying, shouting with tears flowing down her cheeks. She tried to blot them with a tissue she had in her hand.

"Dan's gone. Dan's gone. No. No, it can't be."

We tried to continue the conversation with the parents. One interesting fact I learned was that Johns couldn't swim. This amazed and annoyed me. I couldn't understand how the Navy would allow him aboard ship or in a deploying helicopter squadron without the ability to swim.

After another 20 minutes or so, we ran out of words and left the family with their grief. We explained they'd be hearing further from the Navy, but we wanted to share firsthand what happened.

The father got a grip on himself and thanked us for coming as we were walking out the door.

It was an emotional strain for us all. I decided I didn't ever want to go through that experience again. That day I gained a new respect and appreciation for chaplains and what must be the most difficult part of their job.

On my rescue report, I offered several suggestions which I'm sure others had suggested as well, because they eventually became common practice.

1. All flight deck personnel be supplied and required to wear a life jacket of some kind.
2. In training, remind flight crews that dye marker from vests can be yanked out and thrown in the water to quickly mark a spot.
3. Get better igniting smoke flares.
4. Teach crewmen how to use their clothing as a flotation device.
5. Require all crewmen to be able to swim before deploying.
6. Require that some type of extra flotation device be carried in the helo that could be instantly dropped to someone during an emergency.

Life at the squadron continued with training, test flying, and the usual PIO duties. All was just routine now.

The word's out about a new turbine-powered helicopter built by Kaman called the HU2K. Its gross weight and airspeed are double that of the HUP. The stories we're hearing is that the Kaman helicopter was selected for the Navy to keep them in business even though the helo is underpowered and is having development problems.

There's a new Sikorsky model out that would be perfect to replace the HUP, but the powers that be don't want to hear what the guys in the fleet have to say about operational performance of this helicopter. We suffered through the HUP, yet everyone who mastered it seems to have love and fascination for this unique machine. Some pilots were killed while testing the HU2K. With its single-engine turbine it probably will not have the ability to perform well aboard ship.

My last hop was a training flight up the San Diego coast. I switched from my usual seat as instructor pilot to the left seat as pilot. I skimmed above the surf off Coronado. We then flew over to Pt. Loma looking down on the kelp beds, then flew up the coast off La Jolla under the Miramar Sea Wolf departure corridor and onto Del Mar. It was a gorgeous sunny day in southern California.

On the return it was fun to give brief waves to surfers and pretty girls along the beach as we cruised along at 60 knots about 10 feet above the water. Then I made a couple of landings on the practice pad [no autorotations—no use taking any chances of pranging (causing a crash) anything on the last hop] and a final landing.

During the flight my mind wandered on the challenges of flight training, that very special week in survival school and my buildup in the squadron for the challenge of a long cruise. Memories of all the problems I'd had, had faded and just the pleasant memories remained...and, of course, the rescues.

I know I'll miss the challenges of flying, although I'll still get some in the reserves as a "weekend warrior," but at the same time I'm looking forward to a new adventure in life selling advertising and writing stories for our family-owned community newspapers here in San Diego.

I left the Navy in December 1959, after 2 years in the squadron with 861.7 total flight hours, 634.3 helicopter hours, and 539 shipboard landings plus 62 air and sea rescues. The experience is one of the greatest a young man could ever ask for.

Lieutenant Commander Fortin was right when he said, "Almost every alumnus of the HU-1 school is intensely proud of his long-cruise achievements. He has gained an inner power from the knowledge that he has done a difficult job well. To do the job well is traditional in this squadron. Very few have broken the tradition. He faces the future more confidently—be it in or out of the Navy."

Besides the personal satisfaction and the service to my country, I learned some valuable lessons of life that have served me well. Things like the importance of self-discipline, the right mental attitude, a cheerful spirit, enthusiasm, and a strong faith in God.

Epilogue

THIS BOOK SHARES the adventures of the past.
Today it is important to look to the future. Tactical aircraft will continue to be used in offensive combat operations essential for military and political victory by the forces of the United States and its allies.

Traditionally the "combat SAR" mission took a backseat to the destruction of targets, but today with uncertainty in everything, the retrieving of downed crew members has become the highest priority.

The reasons are obvious and focus on unwanted long-term political liability for limited operations, such as in the case with CWO (Chief Warrant Officer) Durante who was captured in Somalia after a failed raid. The morale of everyone was involved in that incident.

The United States went to great lengths to ensure that the world knew that all hell will break loose if just one downed aviator is captured. All this works to provide the simplest reassurance to our fighting men.

There's no use in going through the exercise of why pilots are not expendable, about the dollars saved when a well-trained pilot is rescued versus the replacement costs, or how pilot morale is affected and how pilot performance is inspired if the pilot knows that the flight is not a suicide mission. It's important that the pilots know that rescue efforts will take place and that someone cares in their time of distress.

The POW faces the threat of brutal treatment in almost all crisis locations in the world today because of the unstable political environments that allow well-armed and irrational groups to hold ultimate power over the groups that are with them or subjected by them. The potential extreme torture and political exploitation make rescue efforts critical and very difficult.

The reassurance that someone cares gives pilots greater confidence in executing their missions and the deeper inner conviction that if downed, they can survive.

So what do we do about planning for rescue in the future?

The helicopter has been an effective rescue vehicle since the Korean War, but it is vulnerable and ultimately speed-limited. At the same time, a dedicated combat SAR force can become a lost art, as it did after the

Korean War, and only a brief reassurance became a reality on an ad hoc basis during the Gulf War.

In the 13 years of SAR experience with the HUP there were no known combat saves. We really were not trained or equipped for such efforts, although many of us made our own preparations and worked with meager checklists. No specialized equipment was developed until the coming of the Vietnam War.

But in Vietnam, SAR efforts relied mostly on the guts and determination of the helicopter crews. Tactical innovations were made, crew members modified equipment, and pilots simply flew into harm's way in spite of the threat and consequences to retrieve downed airmen. Results were seldom great, but the efforts were all heroic.

U.S. Navy SAR helicopters rescued more than 200 downed air crew members from the Gulf of Tonkin and 27 on North Vietnamese territory. The 27 rescues made in North Vietnam were the result of hundreds of attempts and many disastrous failures. Two SAR aircraft and/or helos were lost for every three air crewmen rescued, with one SAR crew member killed for every two rescues. This is not a great record, but it was characterized by overwhelming courage constantly "above and beyond" the call of duty.

HC-7, the designated combat SAR squadron of the Tonkin Gulf, was formed in 1967. Its members received one Medal of Honor, four Navy Crosses, several Silver Stars, and more than 50 Distinguished Flying Crosses. All were awarded for SAR missions. Despite those heroic efforts, only 9 percent of the downed U.S. Navy and Marine aviators, that went down over the beach in North Vietnam, were rescued.

When the Vietnam War began, the single-engine H2 was the standard planeguard for all carrier operations in the Pacific. It had no inherent combat rescue capability, no armor, and no armament except for the pilot's pistol. The lessons had to be relearned in earnest one more time as pilots were being shot down over enemy territory. To go miles into North Vietnam, fly around searching, hovering in position for long minutes, and then egress while being shot at was a terrifying proposition for such a fragile and unreliable machine.

While prospects on the land were problematical, survival in the water was never a sure thing, either. About 30 percent of the aviators who bailed out into the Tonkin Gulf drowned or succumbed to their wounds before being picked up.

Speed is the difference between success and failure in any rescue attempt. In combat a rescue force must be prepositioned as close as possible to the target area.

Time was critical in the North Vietnamese rescues. Seventy-five percent of all aviators captured were grabbed after the first half-hour of being on the ground.

Since the Navy retrieved only about one-sixth of all its fliers known to be downed and alive on the ground in North Vietnam or in adjacent waters, it was obvious that there were enormous factors impeding rescue operations.

Initially it was a lack of good equipment and poor training, but that improved. There was very little that could be done about unfriendly civilian areas over which the pilots had to fly, but they did start egressing over the sparsely populated areas instead of taking the most direct routes.

Intelligence assessments of high-threat areas were also helpful in mission planning, but as the years wore on in North Vietnam, gun positions continued to multiply and eventually the "red-flagged" sights to avoid on the maps turned into large blotches covering square miles.

During the Gulf War, many of the same problems existed, but one key lesson was again affirmed—that was the importance of speed and mobility in making rescues. The so-called golden first half-hour required SAR forces close to the action, but during the pre-ground-war air campaign, it was very difficult to get them there. So pilots with damaged aircraft were sent to remote areas, or to the Persian Gulf itself, where there was an overabundance of friendly shipping.

Those who had to leave the aircraft near the target area were generally captured. The Iraqi terrain was considered similar to a flat, asphalt parking lot, which really provided no place to hide.

During the ground war there were hundreds of capable helicopters that were load-capable, relatively fast, and manned with armed crews, who could all be contacted in minutes over known radio nets, and as a result it expedited finding downed pilots. Army units also provided mobile SAR "hot pads" near the forward line of troops.

Because of the excellent command and control setup, SAR efforts were often coordinated with fixed-wing aircraft, tankers, and supporting armed units in an attempt to insulate the downed pilots from Iraqi ground forces.

In one particular incident, a section of A10 Wart Hogs flew for hours, refueling several times to rendezvous with a rescue helicopter to pick up an F14 crew member who was deep in Iraqi territory.

Navy rescue forces today are better organized, equipped, and trained than in any previous era. Much of this has to do with the capability brought into play by having superior helicopters such as the HH60 Black Hawk.

Pilots can get very innovative with such a powerful machine that can go far and carry a significant load. But advance training is needed for night operations incorporated with multiple censors, command and control via space, real-time intelligence fusion, and other integrating factors that make rescue and high-threat environments possible and survivable.

By the mid-1980s the U.S. Navy was down to one rescue squadron, a reserve unit, the "Protectors" from HC-9, flying old Sikorsky HH-3As. At that time the Navy started to seriously evaluate the feasibility of combat SAR, realizing that as the threat levels continued to rise, it would require more and more offensive action to clear a sanctuary so that SAR activity could commence to retrieve downed pilots.

The result of this evaluation was the creation of a strike rescue mission that would become an integral part of a Carrier Air Wing's mission portfolio. This inspired the need for a more compatible rescue platform and hence the birth of the HH60H "Rescue Hawk," a variant of the multiservice Sikorsky Black Hawk utility helicopter. These new night-capable SAR helos now make up the inventories of two U.S. Navy reserve squadrons.

By early 1993 the USAF started to transfer most of its 13 air rescue service squadrons and detachments from the Air Mobility Command to the Air Combat Command in the Pacific. They normally employ the HH-60H/G "Pave Hawk" helicopter, supported by the HC130H "Night Hercules" tanker. The combat SAR mission for the USAF and the Army has come to overlap somewhat with capabilities of the Special Operations Command (SOC). Much of the SOC organization and resources, together with its tie-in to the national level communications systems, and the fusion of offboard sensor and intelligence information have enabled it to serve in a direct or supportive combat role.

Perhaps this has been the most important step taken since the Gulf War, because of the worldwide playing field of the U.S. Navy and Marines, who still utilize the inventory of this aircraft and crews.

The Navy/Marine long-term future may rest in the V-22 Osprey, which is the high-speed and versatile tilt-rotor aircraft presently (at the time of writing) under cautious development.

Most of today's SAR pilots, however, prefer immediate upgrades to present helicopters, including better and multiband radios, data link, color displays, integrated mission planning systems, standoff defensive/warning systems, night vision devices, forward-looking infrared (FLIR) sensors, helmet sight systems, and aviation night-vision integrated systems (ANVIS).

With the investment and the right tools in training for the future of the combat rescue mission, we move into the precarious world a bit more confidently.

Perhaps the biggest thank you that a SAR pilot could receive from the flying fighter and attack air crews would sound similar to the recognition given to an assembled group of rescue crews during the Gulf War by an A10 squadron commander, "Because we knew you were there, my pilots pressed a little harder, flew a little lower, and dropped our bombs a little better."

About the Author

Dan McKinnon is an expert in commercial aviation travel and survival.

He currently is the owner and president of North American Airlines based at JFK International Airport. North American is a worldwide large jet charter and scheduled airline flying B757 and new generation B737-800 aircraft.

From 1981 to 1985 he served as Chairman of the Civil Aeronautics Board. He was appointed by President Ronald Reagan and oversaw implementation of airline deregulation during the tumultuous period of bankruptcies and adjustment from a government-regulated industry to one controlled by the marketplace.

McKinnon played a key role in U.S. international aviation policy and negotiations for air-route agreements with countries around the world.

On December 31, 1984, McKinnon oversaw the shutdown of the 46-year-old Civil Aeronautics Board in accordance with the President's wishes—the first government regulatory agency ever closed.

McKinnon formerly owned and operated two radio stations in San Diego for 23 years and spent 4 years as publisher of a newspaper in La Jolla, California.

A former Navy pilot, who specialized in rescue efforts and techniques, McKinnon holds the Navy peacetime helicopter rescue record with 62 air-sea saves.

In the middle 1980s he did special projects for the Director of the Central Intelligence Agency.

As part of his military training, and continuously since that time, he has done extensive study in survival, captive, and POW situations.